THE JOY OF LIVING

THE
JOY
OF
LIVING

Unlocking the Secret and Science of Happiness

YONGEY MINGYUR RINPOCHE
WITH ERIC SWANSON

FOREWORD BY DANIEL GOLEMAN

HARMONY BOOKS
NEW YORK

2668301

Published in the United States by Harmony Books,
an imprint of the Crown Publishing Group,
a division of Random House, Inc., New York.
www.crownpublishing.com

HARMONY BOOKS is a registered trademark
and the Harmony Books colophon is a trademark
of Random House, Inc.

Library of Congress Cataloging-in-Publication Data
Yongey Mingyur, Rinpoche, 1975–
The joy of living : unlocking the secret and science of happiness / Yongey Mingyur Rinpoche,
with Eric Swanson ; foreword by Daniel Goleman.
p. cm.
Includes bibliographical references and index.
1. Religious life—Buddhism. I. Swanson, Eric.
II. Goleman, Daniel. III. Title.
BQ5395.Y66 2007
294.3'444—dc22   2006015356

ISBN-13: 978-0-307-34625-4
ISBN-10: 0-307-34625-0

Printed in the United States of America

Book design by Jo Anne Metsch

10 9 8 7 6 5 4 3 2 1

First Edition

CONTENTS

FOREWORD

WE ARE WITNESSING an unparalleled episode in the history of science: a serious, ongoing two-way conversation between scientists and contemplatives. From the scientific perspective, some of this encounter has been sobering. My own branch of science, psychology, had always assumed that its roots were to be found in Europe and America around the start of the twentieth century. That view turns out to be both culture-bound and historically shortsighted: Theories of the mind and its workings—that is, psychological systems—have been developed within most of the great world religions, all from Asia.

Back in the 1970s, traveling in India as a graduate student, I found myself studying Abhidharma, one of the more elegant examples of such an ancient psychology from Buddhism. I was stunned to discover that the basic questions of a science of mind had been explored for millennia, not just a mere century. Clinical psychology, my own field at the time, sought to help alleviate the varieties of emotional pain. But, to my surprise, I found that this millennia-old system articulated a set of methods not just for healing mental suffering, but also for expanding such positive human capacities as compassion and empathy. Yet I had never heard of this psychology anywhere in my own studies.

Today the vigorous dialogue between practitioners of this ancient inner science and modern scientists has blossomed into active collaboration. This working partnership has been catalyzed by the Dalai Lama and the Mind and Life Institute, which for several years has brought together Buddhists and scholars in discussions with modern scientists. What began as exploratory conversations has evolved into a joint research effort. As a result, experts in Buddhist mind science have been working with neuroscientists to design research that will document the neural impact of these varieties of mental training. Yongey Mingyur Rinpoche has been one of the expert practitioners most actively involved in this alliance, working with Richard Davidson, the director of the Waisman Laboratory for Brain Imaging and Behavior at the University of Wisconsin. This research has yielded stunning results, which if replicated will alter forever certain basic scientific assumptions—for example, that systematic training in meditation, when sustained steadily over years, can enhance the human capacity for positive changes in brain activity to an extent undreamed of in modern cognitive neuroscience.

Perhaps the most staggering result to date came in a study of a handful of meditation adepts that included Yongey Mingyur Rinpoche (as he describes in this book). During a meditation on compassion, neural activity in a key center in the brain's system for happiness jumped by 700 to 800 percent! For ordinary subjects in the study, volunteers who had just begun to meditate, that same area increased its activity by a mere 10 to 15 percent. These meditation experts had put in levels of practice typical of Olympic athletes—between ten thousand and fifty-five thousand hours over the course of a lifetime—honing their meditative skills during years of retreat. Yongey Mingyur is something of a prodigy here. As a young boy, he received profound meditation instructions from his father, Tulku Urgyen Rinpoche, one of the most renowned masters to have come out of Tibet, just before the Communist invasion. When he was only thirteen, Yongey Mingyur was inspired to begin a three-year-long meditation retreat. And when he had finished, he was made meditation master of the very next three-year retreat at that hermitage.

Yongey Mingyur is unusual, too, in his keen interest in modern science. He has been an ardent spectator at several of the Mind and Life meetings, and has seized every opportunity to meet one-on-one with scientists who could explain more about their specialties. Many of these conversations have revealed remarkable similarities between key points in Buddhism and modern scientific understanding—not just in psychology, but also with cosmological principles stemming from recent advances in quantum theory. The essence of those conversations is shared in this book.

But these more esoteric points are woven into a larger narrative, a more pragmatic introduction to the basic meditation practices Yongey Mingyur teaches so accessibly. This is, after all, a practical guide, a handbook for transforming life for the better. And that journey begins from wherever we happen to find ourselves, as we take the first step.

Daniel Goleman

THE JOY OF LIVING

THIS BOOK BEGAN life as a relatively simple assignment of splicing together transcripts of Yongey Mingyur Rinpoche's early public lectures at Buddhist centers around the world and editing the whole into a somewhat logically structured manuscript. (It should be noted that "Rinpoche"—which may be roughly translated from the Tibetan as "precious one"—is a title appended to the name of a great master, similar to the way the title "Ph.D." is appended to the name of a person deemed expert in various branches of Western study. According to Tibetan tradition, a master granted the title of Rinpoche is often addressed or referred to by the title alone.)

As is often the case, however, simple tasks tend to assume lives of their own, growing beyond their initial scope into much larger projects. Since most of the transcripts I'd received had been drawn from Yongey Mingyur's early years of teaching, they didn't reflect the detailed understanding of modern science he had acquired through his later discussions with European and North American scientists, his participation in the Mind and Life Institute conferences,[1] and his personal experience as a research subject at the Waisman Laboratory for Brain Imaging and Behavior at the University of Wisconsin, Madison.

Fortunately, an opportunity to work directly with Yongey Mingyur on the manuscript opened up when he took a break from his worldwide teaching schedule for an extended stay in Nepal during the closing months of 2004. I must admit that I was inspired more by dread than by excitement over the prospect of spending several months in a country beset by conflict between government and insurgent factions. But whatever inconveniences I experienced during my stay there were more than offset by the extraordinary chance to spend an hour or two every day in the company of one of the most charismatic, intelligent, and knowledgeable teachers I have ever had the privilege of knowing.

Born in 1975 in Nubri, Nepal, Yongey Mingyur Rinpoche is a rising star among the new generation of Tibetan Buddhist masters trained outside of Tibet. Deeply versed in the practical and philosophical disciplines of an ancient tradition, he is also amazingly conversant in the issues and details of modern culture, having taught for nearly a decade around the world, meeting and speaking with a diverse array of individuals ranging from internationally renowned scientists to suburbanites trying to resolve petty feuds with angry neighbors.

I suspect that the ease with which Rinpoche is able to navigate the complex and sometimes emotionally charged situations he encounters during his worldwide teaching tours stems in part from having been groomed from an early age for the rigors of public life. At the age of three, he was formally recognized by the Sixteenth Karmapa (one of the most highly esteemed Tibetan Buddhist masters of the twentieth century) as the seventh incarnation of Yongey Mingyur Rinpoche, a meditation adept and scholar of the seventeenth century who was noted in particular as a master of advanced methods of Buddhist practice. At about the same time, Dilgo Khyentse Rinpoche informed Rinpoche's parents that their son was also the incarnation of Kyabje Kangyur Rinpoche, a meditation master of immense practical genius—who, as one of the first of the great Tibetan masters to voluntarily accept exile from his homeland in the wake of the political changes that began to shake Tibet in the 1950s, commanded until his death an enormous audience of both Eastern and Western students.

For those unfamiliar with the particulars of the Tibetan system of reincarnation, a note of explanation may be necessary here. According to the Tibetan Buddhist tradition, great teachers who have achieved the highest states of *enlightenment* are said to be moved by enormous compassion to be reborn again and again in order to help all living creatures discover in themselves complete freedom from pain and suffering. The Tibetan term for these passionately committed men and women is *tulku,* a word that may be roughly translated into English as "physical emanation." Undoubtedly the best-known tulku of the modern age is the Dalai Lama, whose current incarnation epitomizes the compassionate commitment to the welfare of others ascribed to a reincarnate master.

Whether you choose to believe that the present Yongey Mingyur Rinpoche has carried the same broad range of practical and intellectual skills through successive incarnations or has mastered them through a truly exceptional degree of personal diligence is up to you. What distinguishes the current Yongey Mingyur Rinpoche from previous holders of the title is the worldwide scale of his influence and renown. While previous incarnations of the Yongey Mingyur Rinpoche line of tulkus has been somewhat limited by the geographical and cultural isolation of Tibet, circumstances have conspired to allow the current holder of the title to communicate the depth and breadth of his mastery to an audience of thousands extending from Malaysia to Manhattan to Monterey.

Titles and pedigrees, however, offer little in the way of protection against personal difficulties, of which Yongey Mingyur Rinpoche certainly has had his share. As he relates with great candor, though raised by a loving family in an area of Nepal noted for its pristine beauty, he spent his early years plagued by what would probably be diagnosed by Western specialists as panic disorder. When he first told me about the depth of anxiety that characterized his childhood, I found it hard to believe that this warm, charming, and charismatic young man had spent much of his childhood in a persistent state of fear. It's a testament not only to his extraordinary strength of character but also to the efficacy of the Tibetan Buddhist practices he presents in this, his first book, that he was able, without recourse to conventional pharmaceutical and therapeutic aid, to master and overcome this affliction.

Rinpoche's personal testimony is not the only evidence of his triumph over devastating emotional pain. In 2002, as one of eight long-term Buddhist meditation practitioners participating in a study conducted by Antoine Lutz, Ph.D., a neuroscientist trained by Francisco Varela, and Richard Davidson, Ph.D., a world-renowned neuroscientist and a member of the Board of Scientific Counselors of the National Institute of Mental Health, Yongey Mingyur underwent a battery of neurological tests at the Waisman Laboratory in Madison, Wisconsin. The tests employed state-of-the-art fMRI technology that—unlike standard MRI technology, which provides only a kind of still photograph of brain/body activity—captures a moment-by-moment pictorial record of changing levels of activity in different areas of the brain. The EEG equipment used to measure the tiny electrical impulses that occur when brain cells communicate was also quite sophisticated. Whereas a typical EEG procedure involves attaching only sixteen electrodes to the scalp in order to measure electrical activity at the surface of the skull, the equipment used in the Waisman lab employed 128 electrodes in order to measure tiny changes in electrical activity deep within the subjects' brains.

The results of both the fMRI and EEG studies of these eight trained meditators were impressive on two levels. While practicing compassion and loving-kindness meditation, the brain area known to be activated in maternal love and empathy was more prominently activated among long-term Buddhist practitioners than among a group of control subjects who had been given meditation instructions one week prior to the scans and asked to practice daily. Yongey Mingyur's capacity to generate such an altruistic and positive state was truly amazing, since even people who don't suffer panic attacks frequently experience feelings of claustrophobia when lying in the narrow space of an fMRI scanner. The fact that he could focus his mind so capably even in a claustrophobic environment suggests that his meditation training overrode his propensity for panic attacks.

More remarkably, the measurements of the long-term practitioners' EEG activity during meditation were apparently so far off the scale of normal EEG readings that—as I understand it—the lab technicians thought at first that the machinery might have been malfunctioning. After hastily double-checking their equipment, though, the technicians were forced to eliminate the possibility of mechanical malfunction and confront the reality that the electrical activity associated with attention and phenomenal awareness transcended anything they'd ever witnessed. In a departure from the typically cautious language of modern scientists, Richard Davidson recalled in a 2005 *Time* magazine interview, "It was exciting. . . . We didn't expect to see anything quite that dramatic."[2]

In the following pages, Yongey Mingyur quite frankly discusses his personal troubles and his struggle to overcome them. He also describes his first childhood meeting with a young Chilean scientist named Francisco Varela, who would eventually become one of the foremost neuroscientists of the twentieth century. Varela was a student of Yongey Mingyur's father, Tulku Urgyen Rinpoche, whose teaching engagements in Europe, North America, and Asia attracted thousands of students. Varela developed a close friendship with Yongey Mingyur, introducing him to Western ideas about the nature and function of the human brain. Seeing his interest in science, others among Tulku Urgyen's Western students began teaching Yongey Mingyur about physics, biology, and cosmology. These early "science lessons," offered to him at the age of nine, had a profound effect on him, eventually inspiring him to find a way to bring together the principles of Tibetan Buddhism and modern science in a manner accessible to those who, unable to wade through scientific publications and either skeptical or overwhelmed by the sheer volume of Buddhist books, nevertheless yearned for a practical means of achieving a lasting sense of personal well-being.

Before Yongey Mingyur could begin such a project, however, he had to complete his formal Buddhist education. Between the ages of eleven and thirteen, he traveled back and forth between his father's hermitage in Nepal and Sherab Ling monastery in India, the primary residence of the Twelfth Tai Situ Rinpoche, one of the most important Tibetan Buddhist teachers alive today. Under the guidance of the Buddhist masters in Nepal and Sherab Ling, he engaged in an intense study of the *sutras,* which represent the actual words of the Buddha, and the *shastras,* a collection of texts that represent Indian Buddhist commentaries on the sutras, as well as seminal texts and commentaries by Tibetan masters. In 1988, at the end of this period, Tai Situ Rinpoche granted him permission to enter the very first three-year retreat program at Sherab Ling.

Created centuries ago in Tibet as the basis of advanced meditation training, the three-year retreat is a highly selective program that involves intensive study of the core techniques of Tibetan Buddhist meditation practice. Yongey Mingyur Rinpoche was one of the youngest students in the known history of Tibetan Buddhism to be al-

lowed to enter this program. His progress during those years was so impressive that after he completed the program, Tai Situ Rinpoche appointed him master of the next retreat at Sherab Ling—making him, at
the age of seventeen, the youngest known retreat master in the history
of Tibetan Buddhism. In his role as retreat master, Yongey Mingyur
completed what amounted to nearly seven years of formal retreat.

In 1994, at the end of the second retreat, he enrolled at a monastic college, known in Tibetan as a *shedra*, to continue his formal education
through intensive study of essential Buddhist texts. The following year, Tai
Situ Rinpoche appointed him principal representative of Sherab Ling,
overseeing the entire range of the monastery's activities and reopening
the shedra there, where he continued his own studies while also serving
as a teacher. For the next several years, Yongey Mingyur Rinpoche divided his time between overseeing the monastery's activities, teaching
and studying at the shedra, and serving as master for yet another three-
year retreat. In 1998, at the age of twenty-three, he took full monastic or-
dination.

Since the age of nineteen—an age when most of us are preoccupied
by more worldly interests—Yongey Mingyur has maintained a grueling
schedule that includes supervising the activities of monasteries in
Nepal and India, worldwide teaching tours, private counseling, com-
mitting to memory hundreds of pages of Buddhist texts, and absorbing
as much as he can from the last living members of the generation of
meditation teachers trained in Tibet.

What has impressed me most during the time I've known him,
though, is his capacity to meet every challenge with not only an envi-
able degree of composure, but also with a sharp, ingeniously timed
sense of humor. On more than one occasion during my stay in Nepal,
while I was droning on over the transcript of our previous day's conversation, he would sometimes pretend to fall asleep or make as if to bolt
out the window. In time I realized he was simply "busting" me for
taking the work too seriously, demonstrating in an especially direct
manner that a certain amount of levity is essential to Buddhist practice. For if, as the Buddha proposed in the first teachings he gave upon
attaining enlightenment, the essence of ordinary life is suffering, then
one of the most effective antidotes is laughter—particularly laughter

at oneself. Every aspect of experience assumes a certain kind of brightness once you learn to laugh at yourself.

This is perhaps the most important lesson Yongey Mingyur offered me during the time I spent with him in Nepal, and I'm as grateful for it as for the profound insights into the nature of the human mind he has been able to offer through his unique ability to synthesize the subtleties of Tibetan Buddhism and the marvelous revelations of modern science. It is my sincerest hope that everyone who reads this book will find his or her own path through the maze of personal pain, discomfort, and despair that characterizes everyday life, and learn, as I did, how to laugh.

As a final note, most of the citations from Tibetan and Sanskrit texts are the work of other translators, true giants in their fields, to whom I owe a large debt of gratitude for their clarity, scholarship, and insight. Those few citations that are not directly attributed to others represent my own translations, carefully worked out with Yongey Mingyur Rinpoche, whose profound understanding of ancient prayers and classic texts offered me a further glimpse into the character of a true Buddhist master.

ERIC SWANSON

PART ONE

THE

GROUND

All sentient beings, including ourselves, already possess the primary cause for enlightenment.

—GAMPOPA, *The Jewel Ornament of Liberation*,
translated by Khenpo Konchog Gyaltsen Rinpoche

1

THE JOURNEY BEGINS

If there is any religion that would cope with modern scientific needs, it would be Buddhism.

—ALBERT EINSTEIN

WHEN YOU'RE TRAINED as a Buddhist, you don't think of Buddhism as a religion. You think of it as a type of science, a method of exploring your own experience through techniques that enable you to examine your actions and reactions in a nonjudgmental way, with the view toward recognizing, "Oh, this is how my mind works. This is what I need to do to experience happiness. This is what I should avoid to avoid unhappiness."

At its heart, Buddhism is very practical. It's about doing things that foster serenity, happiness, and confidence, and avoiding things that provoke anxiety, hopelessness, and fear. The essence of Buddhist practice is not so much an effort at changing your thoughts or your behavior so that you can become a better person, but in realizing that no matter what you might think about the circumstances that define your life, you're already good, whole, and complete. It's about recognizing the inherent potential of your mind. In other words, Buddhism is not so much concerned with getting well as with *recognizing* that you are, right here, right now, as whole, as good, as essentially well as you could ever hope to be.

You don't believe that, do you?

Well, for a long time, neither did I.

I would like to begin by making a confession, which may sound strange coming from someone regarded as a reincarnate lama who is supposed to have done all sorts of wonderful things in previous lifetimes. From earliest childhood, I was haunted by feelings of fear and anxiety. My heart raced and I often broke out in a sweat whenever I was around people I didn't know. There wasn't any reason for the discomfort I experienced. I lived in a beautiful valley, surrounded by a loving family and scores of monks, nuns, and others who were deeply engaged in learning how to awaken inner peace and happiness. Nevertheless, anxiety accompanied me like a shadow.

I was probably about six years old when I first began to experience some relief. Inspired mostly by a child's curiosity, I began climbing into the hills around the valley where I grew up to explore the caves where generations of Buddhist practitioners had spent their lives in meditation. Sometimes I'd go into a cave and pretend to meditate. Of course, I really had no idea how to meditate. I'd just sit there mentally repeating *Om Mani Peme Hung,* a mantra, or repetition of special combinations of ancient syllables, familiar to almost every Tibetan, Buddhist or not. Sometimes I'd sit for hours, mentally reciting the mantra without understanding what I was doing. Nevertheless, I started to feel a sense of calm stealing over me.

Yet even after three years of sitting in caves trying to figure out how to meditate, my anxiety increased until it became what would probably be diagnosed in the West as a full-blown panic disorder. For a while I received some informal instructions from my grandfather, a great meditation master who preferred to keep his accomplishments quiet; but finally I summoned the courage to ask my mother to approach my father, Tulku Urgyen Rinpoche, with my request to study formally with him. My father agreed, and for the next three years he instructed me in various methods of meditation.

I didn't understand much at first. I tried to rest my mind in the way he taught, but my mind wouldn't rest. In fact, during those early years of formal training, I actually found myself growing more distracted than before. All sorts of things annoyed me: physical discomfort, background noises, conflicts with other people. Years later I would come to realize I wasn't actually getting worse; I was simply becoming more *aware* of the constant stream of thoughts and sensations I'd never recognized before. Having watched other people go through the same process, I realize now that this is a common experience for people who are just learning how to examine their mind through meditation.

Although I did begin to experience brief moments of calmness, dread and fear continued to haunt me like hungry ghosts—especially since every few months I was sent to Sherab Ling monastery in India (the primary residence of the Twelfth Tai Situ Rinpoche, one of the greatest masters of Tibetan Buddhism alive today, and one of my most influential teachers, whose great wisdom and kindness in guiding my own development are debts I can never repay) to study under new teachers among unfamiliar students, and then sent back to Nepal to continue training under my father. I spent almost three years that way, shuttling back and forth between India and Nepal, receiving formal instruction from my father and from my teachers at Sherab Ling.

One of the most terrible moments came shortly before my twelfth birthday, when I was sent to Sherab Ling for a special purpose, one that I had been dreading for a long time: formal enthronement as the incarnation of the First Yongey Mingyur Rinpoche. Hundreds of people attended the ceremony, and I spent hours accepting their gifts and giving them blessings as if I were somebody really important instead of just a terrified twelve-year-old boy. As the hours passed, I turned so pale that my older brother, Tsoknyi Rinpoche, who was standing beside me, thought I was going to faint.

When I look back on this period, and on all the kindness that was shown to me by teachers, I wonder how I ever could have felt as fearful as I did. In hindsight, I can see that the basis of my anxiety lay in the fact that I hadn't truly recognized the real nature of my mind. I had a basic intellectual understanding, but not the kind of direct experience that would have enabled me to see that whatever terror or discomfort I felt was a product of my own mind, and that the unshakable basis of serenity, confidence, and happiness was closer to me than my own eyes.

At the same time that I began my formal Buddhist training, something wonderful was taking place; though I didn't realize it at the time, this new turn of events would have a lasting impact on my life and actually accelerate my personal progress. I was gradually being introduced to the ideas and discoveries of modern science—in particular the study of the nature and function of the brain.

A MEETING OF MINDS

We have to go through the process of sitting down and examining the mind and examining our experience to see what is really going on.
    —KALU RINPOCHE, *The Gem Ornament of Manifest Instructions,*
        edited by Caroline M. Parke and Nancy J. Clarke

I was only a child when I met Francisco Varela, a Chilean biologist who would later become one of the most renowned neuroscientists of the twentieth century. Francisco had come to Nepal to study the Buddhist method of mental examination and training under my father, whose reputation had attracted quite a number of Western students. When we weren't studying or practicing, Francisco would often talk to me about modern science, especially his own specialty, the structure and function of the brain. Of course, he was careful to frame his lessons in terms a nine-year-old boy could understand. As others among my father's Western students recognized my interest in science, they too began teaching me what they knew of modern theories about biology, psychology, chemistry, and physics. It was a little bit like learning two languages at the same time: Buddhism on the one hand, modern science on the other.

I remember thinking even then that there didn't seem to be much difference between the two. The words were different, but the meaning seemed pretty much the same. After a while, I also began to see that the ways in which Western and Buddhist scientists approached their subjects were remarkably alike. Classical Buddhist texts begin by presenting a theoretical or philosophical basis of examination, commonly referred to as the "Ground." They then move on to various methods of practice, commonly referred to as the "Path," and finally conclude with an analysis of the results of personal experiments and suggestions for further study, typically described as the "Fruit." Western scientific investigation often follows a similar structure, beginning with a theory or hypothesis, an explanation of the methods through which the theory is tested, and an analysis comparing the results of the experiments against the original hypothesis.

What fascinated me most about simultaneously learning about modern science and Buddhist practice was that while the Buddhist approach was able to teach people an introspective or subjective method for realizing their full capacity for happiness, the Western perspective explained in a more objective fashion why and how the teachings worked. By themselves, Buddhism and modern science both provided extraordinary insights into the workings of the human mind. Taken together, they formed a more complete and intelligible whole.

Near the end of that period of traveling between India and Nepal, I learned that a three-year retreat program was about to begin at Sherab Ling monastery. The master of the retreat would be Saljay Rinpoche, one of my principal teachers at Sherab Ling. Saljay Rinpoche was considered one of the most accomplished masters of Tibetan Buddhism of his day. A gentle man with a low voice, he had an amazing ability to do or say exactly the right thing at exactly the right time. I'm sure some of you have spent time around people who had a similar kind of impact, people able to teach incredibly profound lessons without appearing to be teaching at all. Just the way they *are* is a lesson that lasts for the rest of your life.

Because Saljay Rinpoche was very old, and this would most likely be the last retreat he might ever lead, I wanted very much to take part in it. I was only thirteen years old, however, an age generally considered too young to tolerate the rigors of three years in retreat. But I begged my father to intervene on my behalf, and in the end, Tai Situ Rinpoche granted me permission to participate.

Before I describe my experiences during those three years, I feel it's necessary to take some time to speak a little bit about the history of Tibetan Buddhism, which I think may help to explain why I was so eager to enter the retreat.

THE IMPORTANCE OF LINEAGE

Conceptual knowledge is not enough . . . you must have the conviction that comes from personal experience.

—THE NINTH GYALWANG KARMAPA,
Mahāmudrā: The Ocean of Definitive Meaning,
translated by Elizabeth M. Callahan

The method of exploring and working directly with the mind that we call Buddhism has its source in the teachings of a young Indian nobleman named Siddhartha. Upon witnessing firsthand the terrible misery experienced by people who had not grown up in the same privileged environment he enjoyed, Siddhartha gave up the security and comforts of his home to find a solution to the problem of human suffering. Suffering takes many forms, ranging from the nagging whisper that we would be happier "if only" some small aspect of our lives were different, to the pain of illness and the terror of death.

Siddhartha became an ascetic, wandering across India to study under teachers who professed to have found the solution he was seeking. Unfortunately, none of the answers they provided and none of the practices they taught him seemed entirely complete. At last he decided to abandon outside advice altogether and seek the solution to the problem of suffering in the place he had begun to suspect it originated: within his own mind. In a place called Bodhgaya, in the northeastern Indian province of Bihar, he sat under the shelter of a tree and sank deeper and deeper into his own mind, determined to find the answers he sought, or die in the attempt. After many days and nights he finally discovered what he was looking for: a fundamental awareness that was unchanging, indestructible, and infinite in scope. When he emerged from this state of profound meditation, he was no longer Siddhartha. He was the Buddha, a Sanskrit title that means "the one who is awake."

What he had awakened to was the full potential of his own nature, which had previously been limited by what is commonly referred to as dualism—the idea of a distinct and inherently real "self" that is separate from an apparently distinct and inherently real "other." As we'll explore later, dualism is not a "character flaw" or defect. It's a complex survival mechanism deeply rooted in the structure and function of the brain—which, along with other mechanisms, can be changed through experience.

The Buddha recognized this capacity for change through introspective examination. The ways in which mistaken concepts become embedded in the mind and the means for cutting through them were the subjects of the teaching he gave over the next forty years of his life as he traveled throughout India, attracting hundreds, perhaps thousands, of students. More than 2,500 years later, modern scientists are beginning to demonstrate through rigorous clinical research that the insights he'd gained through subjective examination are amazingly accurate.

Because the scope of the Buddha's insight and perception extended far beyond the ordinary ideas people hold about themselves and about the nature of reality, he was compelled—like other great teachers before and after him—to communicate what he'd learned through parables, examples, riddles, and metaphors. He had to use words. And though these words were eventually written down in Sanskrit, Pali, and other languages, they've always been handed down orally, generation after generation. Why? Because when we hear the words of the Buddha and of the masters who followed him and achieved the same freedom, we have to *think* about their meaning and *apply* that meaning to our own lives. And when we do this, we generate changes in the structure and functions of our brains, many of which will be discussed in the following pages, creating for ourselves the same freedom the Buddha experienced.

In the centuries following the Buddha's death, his teachings began spreading to many countries, including Tibet, whose geographical isolation from the rest of the world provided a perfect setting for successive generations of students and teachers to devote themselves exclusively to study and practice. The Tibetan masters who achieved enlightenment and became Buddhas in their own lifetimes would then pass on everything they had learned to their most promising students, who, in their turn, passed this wisdom on to their own students. In this way, an unbroken lineage of teaching based upon the instructions of the Buddha, as faithfully recorded by his early followers, and on the detailed commentaries on those original teachings, was established in Tibet. But the real power of the lineage of Tibetan Buddhism, what gives it such purity and strength, is the direct connection between the hearts and minds of the masters who passed the core teachings of the lineage orally, and often secretly, to their students.

Because many areas of Tibet are themselves isolated from each other by mountains, rivers, and valleys, it was often difficult for masters and students to travel around, sharing what they'd learned with one another. As a result, the teaching lineages in different regions evolved in slightly different ways. There are currently four major schools, or lineages, of Tibetan Buddhism: Nyingma, Sakya, Kagyu, and Gelug. Although each of these major schools developed at different times and in different areas of Tibet, they share the same basic principles, practices, and beliefs. The differences between them, similar to the distinctions, I'm told, that exist between various denominations of Protestantism, lie mainly in terminology and often quite subtle approaches to scholarship and practice.

The oldest of these lineages, established between the seventh and early ninth centuries C.E., when Tibet was ruled by kings, is the Nyingma school—*nyingma* being a Tibetan term that may be roughly translated as "the old ones." Sadly, the last of the Tibetan kings, Langdarma—for political and personal reasons—initiated a violent repression of Buddhism. Although Langdarma ruled for only four years before he was assassinated in 842 C.E., for nearly 150 years after his death the early lineage of Buddhist teachings remained a kind of "underground" movement, as Tibet underwent massive political changes, eventually reforming itself into a series of separate but loosely federated feudal kingdoms.

These political changes provided an opportunity for Buddhism to slowly and quietly reassert its influence, as Indian teachers traveled to Tibet and interested students made the arduous trek across the Himalayas to study directly under Indian Buddhist masters. Among the first schools to take root in Tibet during this period was the Kagyu order, which takes its name from the Tibetan terms *ka*, roughly translated into English as "speech," or "instruction," and *gyu*, a Tibetan term essentially meaning "lineage." The basis of the Kagyu school lies in the tradition of passing instructions orally from master to student, preserving in this way an almost unparalleled purity of transmission.

The Kagyu tradition originated in India during the tenth century C.E., when an extraordinary man named Tilopa awoke to his full potential. Over several generations the insights Tilopa had achieved, and the practices through which he achieved them, were passed from master to student, eventually reaching Gampopa, a brilliant Tibetan who'd given up his practice as a doctor to pursue the teachings of the Buddha. Gampopa transmitted everything he'd learned to four of his most promising students, who established their own schools in different areas of Tibet.

One of these students, Dusum Khyenpa (a Tibetan name that may be translated as "the seer of the three times"—the past, the present, and the future), founded what is today known as the Karma Kagyu lineage, which derives its name from the Sanskrit word *karma,* which may be roughly translated as "action" or "activity." In the Karma Kagyu tradition, the entire set of teachings, representing more than a hundred volumes' worth of philosophical and practical instruction, is transmitted orally by the master of the lineage, known as the Karmapa, to a handful of students—several of whom incarnate through succeeding generations specifically in order to transmit the entirety of the teachings to the next incarnation of the Karmapa—in order to preserve and protect these incalculable lessons in the pure form in which they were delivered more than a thousand years ago.

There's no equivalent in Western culture for this kind of direct and continuous transmission. The closest we can come to imagining how it might work is to think of someone like Albert Einstein approaching his most able students and saying, "Excuse me, but I'm now going to dump everything I've ever learned into your brain. You keep it for a while, and when I come back in another body twenty or thirty years from now, your job is to dump everything I've taught you back into the brain of some youngster you'll only be able to recognize as me through the insights I'm passing on to you. Oh, and by the way, just in case anything goes awry, you'll need to pass everything I'm now going to teach you to a few other students whose qualities you'll be able to recognize on the basis of what I'm about to show you—just to make sure that nothing gets lost."

Before he passed away in 1981, the Sixteenth Karmapa transmitted this precious body of teachings to several of his main students, known as his "Heart Sons," and charged them with transmitting it to the next incarnation of the Karmapa, while at the same time ensuring that they were preserved intact by passing them on in their entirety to other exceptional students. One of the most prominent of the Sixteenth Karmapa's Heart Sons, the Twelfth Tai Situ Rinpoche, considered me a promising student and facilitated my travel to India to study under the masters assembled at Sherab Ling monastery.

As I mentioned earlier, the distinctions between the different lineages are very small, usually only involving minor variations in terminology and approaches to study. For example, in the Nyingma lineage—of which my father and several of my later teachers were regarded as especially accomplished masters—the teachings on the fundamental nature of the mind are referred to by the term *dzogchen,* a Tibetan word meaning "great perfection." In the Kagyu tradition, the lineage in which Tai Situ Rinpoche, Saljay Rinpoche, and many of the teachers assembled at Sherab Ling were primarily trained, the teachings on the essence of mind are collectively referred to as *mahamudra,* a word that may be roughly translated as "great seal." There is very little difference between the two sets of teachings, except perhaps that the dzogchen teachings focus on cultivating a deep understanding of the *view* of the fundamental nature of mind, while the mahamudra teachings tend to focus on *meditation practices* that facilitate direct experience of the nature of mind.

In the modern world of airplanes, automobiles, and telephones, it's a lot easier for teachers and students to travel around, so whatever differences may have developed with different schools in the past have become less significant. One thing that hasn't changed, however, is the importance of receiving direct transmission of the teachings from those who have mastered them. Through direct connection with a living master, something incredibly precious is transferred; as if some living, breathing thing is passed from the heart of the master into that of the student. It is in this direct way that the teachings given during the three-year retreat are passed from master to student, which may perhaps explain why I was so eager to enter the retreat program at Sherab Ling.

MEETING MY MIND

Just realizing the meaning of mind encompasses all understanding.
—JAMGÖN KONGTRUL, *Outline of Essential Points,*
translated by Maria Montenegro

I'd like to say that everything got better once I was safely settled among the other participants in the three-year retreat at Sherab Ling. On the contrary, however, my first year in retreat was one of the worst of my life. All the symptoms of anxiety I'd ever experienced—physical tension, tightness in the throat, dizziness, and waves of panic that were especially intense during group practices—attacked in full force. In Western terms, I was having a nervous breakdown.

In hindsight, I can say that what I was actually going through was what I like to call a "nervous breakthrough." Completely cut off from the distractions of everyday life, I found myself in the position of having to directly confront my own mind—which at that point was not a very pleasant bit of scenery to stare at day after day. With each passing week it seemed that the mental and emotional landscape I was looking at grew more and more frightening. Finally, as that first year of retreat came to a close, I found myself having to make a choice between spending the next two years hiding in my room or accepting the full truth of the lessons I'd learned from my father and other teachers: that whatever problems I was experiencing were habits of thought and perception ingrained in my own mind.

I decided to follow what I'd been taught.

For three days I stayed in my room meditating, using many of the techniques described later in this book. Gradually I began to recognize how feeble and transitory the thoughts and emotions that had troubled me for years actually were, and how fixating on small problems had turned them into big ones. Just by sitting quietly and observing how rapidly, and in many ways illogically, my thoughts and emotions came and went, I began to recognize in a direct way that they weren't nearly as solid or real as they appeared to be. And once I began to let go of my belief in the story they seemed to tell, I began to see the "author" beyond them—the infinitely vast, infinitely open awareness that is the nature of mind itself.

Any attempt to capture the direct experience of the nature of mind in words is impossible. The best that can be said is that the experience is immeasurably peaceful, and, once stabilized through repeated experience, virtually unshakable. It's an experience of absolute well-being that radiates through all physical, emotional, and mental states—even those that might be ordinarily labeled as unpleasant. This sense of well-being, regardless of the fluctuation of outer and inner experiences, is one of the clearest ways to understand what Buddhists mean by "happiness," and I was fortunate to have caught a glimpse of it during my three days of isolation.

At the end of those three days, I left my room and rejoined the group practices. It took about two more weeks of concentrated practice to conquer the anxiety that had accompanied me throughout my childhood, and to realize through direct experience the truth of what I'd been taught. From that point on, I haven't experienced a single panic attack. The sense of peace, confidence, and well-being that resulted from this experience—even under conditions that might objectively be regarded as stressful—has never wavered. I take no personal credit for this transformation in my experience, because it has only come about through making the effort to apply directly the truth handed down by those who'd preceded me.

I was sixteen years old when I came out of retreat, and much to my surprise, Tai Situ Rinpoche appointed me master of the very next retreat, which was to commence almost immediately. So, within a few months, I found myself back in the retreat house, teaching the preliminary and advanced practices of the Kagyu lineage, providing the new retreat participants access to the same line of direct transmission I'd received. Even though I was now the retreat master, from my point of view it was a wonderful opportunity to spend nearly seven continuous years of intensive retreat practice. And this time I didn't spend a single moment cowering in fear in my own little room.

As the second retreat came to a close, I enrolled for one year in Dzongsar monastic college, quite near to Sherab Ling. The idea was suggested by my father, and Tai Situ Rinpoche readily agreed. Under the direct guidance of the head of the college—Khenchen Kunga Wangchuk, a great scholar who had only recently arrived in India from Tibet—I had the great good fortune to further my education in the philosophical and scientific disciplines of Buddhism.

The method of study at a traditional monastic college is quite different from that at most Western universities. You don't get to choose your classes or sit in a nice classroom or lecture hall, listening to professors give their opinions and explanations of particular subjects, or write essays and take written exams. In a monastic college you're required to study a vast number of Buddhist texts, and almost every day there are "pop quizzes" in which a student whose name is pulled from a jar is required to give a spontaneous commentary on the meaning of a specific section of a text. Our "exams" consisted sometimes of composing written commentaries on the texts we'd studied, and sometimes of public debates in which the teachers would point unexpectedly to individual students, challenging them to provide precise answers to unpredictable questions on the fine points of Buddhist philosophy.

At the end of my first year as a student at Dzongsar, Tai Situ Rinpoche embarked on a series of worldwide teaching tours and assigned me the task of overseeing, under his direction, the day-to-day activities of Sherab Ling as well as the responsibility for reopening the shedra on the monastery grounds, studying and working as an assistant teacher there. He also charged me with leading the next several three-year retreats at Sherab Ling. Since I owed so much to him, I didn't hesitate to accept these responsibilities. If he trusted me to carry out these duties, who was I to question his decision? And, of course, I was fortunate enough to live in an age when I could always count on telephone calls to receive his direct guidance and direction.

Four years passed in this way, overseeing the affairs of Sherab Ling, completing my education and teaching at the new shedra, and giving direct transmissions to the students in retreat. Toward the end of those four years, I traveled to Bhutan to receive from Nyoshul Khen

Rinpoche, a dzogchen master of extraordinary insight, experience, and ability, direct transmission of oral teachings known as *Trekchö* and *Tögal*—which might be translated roughly as "primordial purity" and "spontaneous presence." These teachings are given to only one student at a time, and I was, to say the least, overwhelmed at being chosen to receive this direct transmission, and cannot help but consider Nyoshul Khen Rinpoche, alongside Tai Situ Rinpoche, Saljay Rinpoche, and my father, to have been one of the most influential teachers in my life.

The opportunity to receive these transmissions also taught me, in an indirect way, the extremely valuable lesson that to whatever degree a person commits himself or herself to the welfare of others, he or she is repaid a thousandfold by opportunities for learning and advancement. Every kind word, every smile you offer someone who might be having a bad day, comes back to you in ways you'd never expect. How and why this occurs is a subject that we'll examine later on, since the explanation has a lot to do with the principles of biology and physics I learned about once I'd begun traveling around the world and working more directly with the masters of modern science.

LIGHT FROM THE WEST

> *One single torch can dissipate the accumulated darkness*
> *of a thousand eons.*
>
> —TILOPA, *Mahāmudrā of the Ganges,*
> translated by Maria Montenegro

Because my schedule during the years following my first retreat was rather full, I didn't have much time to follow the advances taking place in neuroscience and related fields of cognitive research, or to digest the discoveries in physics that had entered the mainstream. In 1998, however, my life took an unexpected turn when my brother, Tsoknyi Rinpoche, who'd been scheduled to teach in North America, couldn't make the trip and I was sent in his place. It was my first long visit to the West. I was twenty-three years old. Though I didn't know it as I boarded the plane to New York, the people I would meet during this tour would shape the direction of my thinking for many years to come.

Giving generously of their time and providing me with a mountain of books, articles, DVDs, and videotapes, they introduced me to the ideas of modern physics and the latest developments in neuroscientific, cognitive, and behavioral research. I was very excited, because the scientific research aimed at studying the effects of Buddhist training had become so rich and detailed—and, most important, comprehensible to people like me who weren't trained scientists. And since my knowledge of the English language wasn't advanced at that point, I'm doubly grateful to the people who took so much time to explain the information in terms I could understand. For example, there are no equivalent words in Tibetan for terms like "cell," "neuron," or "DNA"—and the verbal somersaults people had to go through to help me comprehend such things were so complicated that we almost always ended up in fits of laughter.

While I'd been busy with my studies in and out of retreat, my friend Francisco Varela had been working with the Dalai Lama to organize dialogues between modern scientists and Buddhist monks and scholars. Those dialogues evolved into the Mind and Life Institute conferences, during which experts in various fields of modern science and Buddhist studies came together to exchange ideas on the nature and workings of the mind. I was lucky enough to be able to attend the conference in Dharamsala, India, in March 2000, and the conference at MIT in Cambridge, Massachusetts, in 2003.

I learned a lot about the biological mechanisms of the mind during the Dharamsala conference. But it was the MIT conference—which focused on the correlations between the introspective Buddhist methods of exploring experience and the objective approach of modern science—that got me thinking about how to bring what I'd learned during my years of training to people who weren't necessarily familiar with Buddhist practice or the intricacies of modern science.

In fact, as the MIT conference progressed, a question began to emerge: What would happen if the Buddhist and Western approaches were combined? What could be learned by bringing together information provided by individuals trained to offer detailed subjective descriptions of their experiences and the objective data provided by machines capable of measuring minute changes in the activity of the brain? What facts might the introspective methods of Buddhist practice provide that Western lines of technological research cannot? What insights might the objective observations of clinical research be able to offer to Buddhist practitioners?

As the conference ended, participants from both the Buddhist and the Western scientific panels recognized not only that both sides stood to gain enormously through finding ways to work together, but also that the collaboration itself represented a major opportunity to improve the quality of human life. In his closing remarks, Eric S. Lander, Ph.D., a professor of molecular biology at MIT and the director of the Whitehead Institute/MIT Center for Genome Research, pointed out that while Buddhist practices emphasize attaining increased levels of mental awareness, the focus of modern science has rested on refining ways to restore mentally ill patients to a state of normalcy.

"Why stop there?" he asked the audience. "Why are we satisfied with saying we're not mentally ill? Why not focus on getting better and better?"

Professor Lander's questions set me thinking about creating some way to offer people an opportunity to apply the lessons of Buddhism and modern science to the problems they face in their everyday lives. As I learned the hard way during my first year in retreat, theoretical understanding alone is simply not enough to overcome the psychological and biological habits that create so much heartache and pain in daily life. For real transformation to occur, theory has to be applied through practice.

I'm extremely grateful to the Buddhist teachers who provided me, during my early years of training, with such profound philosophical insight and the practical means of applying it. But I'm equally obliged to the scientists who have given so generously of their time and effort, not only toward reevaluating and rephrasing everything I learned in terms that are perhaps more easily accessible to Westerners, but also toward validating the results of Buddhist practice through extensive laboratory research.

How lucky we are to be alive at this unique moment in human history, when the collaboration between Western and Buddhist scientists is poised to offer all humanity the possibility of achieving a level of well-being that defies imagination! My hope in writing this book is that everyone who reads it will recognize the practical benefits of applying the lessons of this extraordinary collaboration, and that they will realize for themselves the promise of their full human potential.

2

THE INNER SYMPHONY

A collection of parts produces the concept of a vehicle.
—*Samyuttanikāya,*
translated by Maria Montenegro

ONE OF THE first lessons I learned as a Buddhist was that every sentient being—that is, every creature endowed with even a very basic sense of awareness—can be defined by three basic aspects or characteristics: body, speech, and mind. *Body,* of course, refers to the physical part of our being, which is constantly changing. It's born, grows up, gets sick, ages, and eventually dies. *Speech* refers not only to our ability to talk, but also to all the different signals we exchange in the form of sounds, words, gestures, and facial expressions, and even the production of pheromones, which are chemical compounds secreted by mammals that subtly influence the behavior and development of other mammals. Just like the body, speech is an impermanent aspect of experience. All the messages we exchange through words and other signs come and go in their time. And when the body dies, the capacity of speech dies with it.

Mind is harder to describe. It's not a "thing" we can point to as easily as we can identify the body or speech. However deeply we investigate this aspect of being, we can't really locate any definite object that we can call the mind. Hundreds, if not thousands, of books and articles have been written in an attempt to describe this elusive aspect of being. Yet in spite of all the time and effort spent on trying to identify what and where the mind is, no Buddhist—and no Western scientist, for that matter—has been able to say once and for all, "Aha! I found the mind! It's located in this part of the body, it looks like this, and this is how it works."

At best, centuries of investigation have been able to determine that the mind has no specific location, shape, form, color, or any other tangible quality we can ascribe to other basic aspects such as the location of the heart or lungs, the principles of circulation, and the areas that control essential functions like the regulation of metabolism. How much easier it would be to say that something so frustratingly indefinable as the mind doesn't exist at all! How much easier it would be to dispatch the mind to the realm of imaginary things like ghosts, goblins, and fairies!

But how could anyone realistically deny the existence of the mind? We think. We feel. We recognize when our backs hurt or our feet fall asleep. We know when we're tired or alert, happy or sad. The inability to precisely locate or define a phenomenon doesn't mean that it doesn't exist. All it really means is that we haven't yet accumulated sufficient information to propose a workable model. To use a simple analogy, you might compare the scientific understanding of the mind to your own acceptance of something as simple as the power of electricity. Flipping a light switch or turning on a TV doesn't require a detailed understanding of circuitry or electromagnetic energy. If the light doesn't work, you replace the bulb. If the TV doesn't work, you check the cable or satellite connection. You may have to replace a burned-out lightbulb, tighten your connection between the TV and the wires connecting it to your cable box or satellite dish, or replace a blown fuse. At worst, you may have to call a technician. But underlying all these actions is a basic understanding, or faith, that electricity *works*.

A similar situation underlies the operation of the mind. Modern science has been able to identify many of the cellular structures and processes that contribute to the intellectual, emotional, and sensory events that we associate with mental functioning. But it has yet to identify anything close to what constitutes "the mind" itself. In fact, the more precisely scientists scrutinize mental activity, the more closely they approach the Buddhist understanding of mind as a perpetually evolving *event* rather than a distinct entity.

Early translations of Buddhist texts attempted to identify the mind as a distinct sort of "thing" or "stuff" that exists beyond the limits of current scientific comprehension. But those incorrect translations were based on early Western assumptions that all experience might eventually be related to some aspect of physical function. More recent interpretations of classical texts reveal an understanding much closer to the modern scientific conception of "the mind" as a kind of constantly evolving *occurrence* arising through the interaction of neurological habits and the unpredictable elements of immediate experience.

Buddhists and modern scientists agree that having a mind is what sets all sentient, or conscious, beings apart from other organisms such as grass or trees—and certainly from things that we wouldn't necessarily consider alive, like rocks, candy wrappers, or blocks of cement. The mind, in essence, is the most important aspect of all creatures that share the attribute of being sentient. Even an earthworm has a mind. Granted, it may not be as sophisticated as the human mind; but then again, there may be some virtue in simplicity. I've never yet heard of an earthworm that stayed up all night worrying about the stock market.

Another issue on which Buddhists and most modern scientists agree is that the mind is the most important aspect of a sentient being's nature. The mind is, in a sense, the puppet master, while the body and the various forms of communication that constitute "speech" are merely its puppets.

You can test this idea of the mind's role for yourself. If you scratch your nose, what is it that recognizes an itch? Is the body, of itself, able to recognize itching? Does the body direct itself to raise its hand and scratch its nose? Is the body even capable of making the distinction between the itch, the hand, and the nose? Or take the example of thirst. If you're thirsty, it's the mind that first recognizes thirst, urges you to ask for a glass of water, directs your hand to accept the glass and bring it to your mouth, and then tells you to swallow. It's the mind that then registers the pleasure of satisfying a physical need.

Even though we can't see it, the mind is always present and active. It's the source of our own ability to recognize the difference between a building and a tree, between rain and snow, between a clear sky and a cloud-filled one. But because having a mind is such a basic condition of our experience, most of us take it for granted. We don't bother to ask ourselves what it is that thinks, *I want to eat; I want to go; I want to sit.* We don't ask ourselves, "Is the mind inside the body or beyond it? Does it start somewhere, exist somewhere, and stop somewhere? Does it have a shape or a color? Does it even exist at all, or is it just the random activity of brain cells that have, over time, accumulated the force of habit?" But if we want to cut through all the varieties and levels of pain, suffering, and discomfort we experience in daily life and grasp the full significance of having a mind, we have to make some attempt to look at the mind and distinguish its main features.

The process is actually very simple. It only seems difficult at first because we're so used to looking at the world "out there," a world that seems to be so full of interesting objects and experiences. When we look at our mind, it's like trying to see the back of our head without the aid of a mirror.

So now I'd like to propose a simple test to demonstrate the problem of trying to look at the mind according to our normal way of understanding things. Don't worry. You can't fail this test, and you don't need a Number 2 pencil to fill out any forms.

Here's the test: The next time you sit down to lunch or dinner, ask yourself, "What is it that thinks that this food tastes good—or not so good? What is it that recognizes eating?" The immediate answer seems obvious: "My brain." But when we actually take a look at the brain from the perspective of modern science, we find that the answer isn't quite so simple.

WHAT'S GOING ON IN THERE?

> *All phenomena are projections of the mind.*
>
> —THE THIRD GYALWANG KARMAPA, *Wishes of Mahāmudrā,*
> translated by Maria Montenegro

If all we want is to be happy, why do we need to understand anything about the brain? Can't we just think happy thoughts, imagine our bodies filled with white light, or fill our walls with pictures of bunnies and rainbows and leave it at that? Well . . . maybe.

Unfortunately, one of the main obstacles we face when we try to examine the mind is a deep-seated and often unconscious conviction that "we're born the way we are and nothing we can do can change that." I experienced this same sense of pessimistic futility during my own childhood, and I've seen it reflected again and again in my work with people around the world. Without even consciously thinking about it, the idea that we can't alter our minds blocks our every attempt to try.

People I've spoken with who try to make a change using affirmations, prayers, or visualizations admit that they often give up after a few days or weeks because they don't see any immediate results. When their prayers and affirmations don't work, they dismiss the whole idea of working with the mind as a marketing gimmick designed to sell books.

One of the nice things about teaching around the world in the robes of a Buddhist monk and with an impressive title is that people who wouldn't usually give an ordinary person the time of day are very happy to talk to me as if I were somebody important enough to take seriously. And during my conversations with scientists around the world, I've been amazed to see that there is a nearly universal consensus in the scientific community that the brain is structured in a way that actually does make it possible to effect real changes in everyday experience.

Over the past ten years or so, I've heard a lot of very interesting ideas from the neuroscientists, biologists, and psychologists with whom I've spoken. Some of what they've said has challenged ideas I was brought up with; other things have confirmed what I'd been taught, though from a different perspective. Whether we've agreed or not, the most valuable thing I've learned from these conversations is that taking the time to gain even a partial understanding of the structure and function of the brain provides a more grounded basis for understanding from a scientific perspective how and why the techniques I learned as a Buddhist actually work.

One of the most interesting metaphors about the brain I've come across was a statement made by Robert B. Livingston, M.D., founding chairman of the Department of Neurosciences at the University of California, San Diego. During the first Mind and Life Institute conference, in 1987, Dr. Livingston compared the brain to "a symphony, well tuned and well disciplined."[1] Like a symphony orchestra, he explained, the brain is made up of groups of players that work together to produce particular results, such as movements, thoughts, feelings, memories, and physical sensations. Although these results may appear fairly simple when you watch someone yawn, blink, sneeze, or raise an arm, the sheer number of players involved in such simple actions, and the range of interactions among them, form an amazingly complex picture.

To better understand what Dr. Livingston was saying, I had to ask people to help me understand the information in the mountain of books, magazines, and other materials I'd received during my first few tours of the West. A lot of the material was extremely technical, and as I tried to understand it all, I found myself feeling a huge amount of compassion for aspiring scientists and medical students.

Fortunately, I've been able to talk at length with people more knowledgeable than I am in such areas, who translated all the scientific jargon into simple terms that I could understand. I hope the time and effort they expended was as helpful to them as it was to me. Not only did my English vocabulary increase enormously, but I also gained an understanding of how the brain works in a way that made very simple "people sense." And as my grasp of the essential details improved, it became clearer to me that for someone who was not raised in the Buddhist tradition, a basic appreciation of the nature and the role of the "players" that Dr. Livingston spoke about is essential to understanding how and why the Buddhist techniques of meditation actually work on a purely physiological level.

I was also fascinated to learn from a scientific point of view what had happened inside my own brain that enabled me to go from being a panic-stricken child to someone who can travel around the world and sit without any trace of fear in front of hundreds of people who've come to hear me teach. I can't really explain why I'm so curious about understanding the physical reasons behind the changes that occur after years of practice, while so many of my teachers and contemporaries are satisfied with the shift in consciousness itself. Maybe in a former life I was a mechanic.

But, getting back to the brain: In very basic "people terms," most brain activity seems due to a very special class of cells called *neurons.* Neurons are very social cells: They love to gossip. In some ways they're like naughty schoolchildren constantly passing notes and whispering to one another—except that the secret conversations between neurons are mainly about sensations, movement, solving problems, creating memories, and producing thoughts and emotions.

These gossipy cells look a lot like trees, made up of a trunk, known as an *axon,* and branches reaching out to send and receive messages to and from other branches and other nerve cells running through the muscle and skin tissues, vital organs, and sense organs. They pass their messages to one another across tiny gaps between the closest branches. These gaps are called *synapses.* The actual messages that flow across these gaps are carried in the form of chemical molecules called *neurotransmitters,* which create electrical signals that can be measured by an EEG. Some of these neurotransmitters are pretty well known to people nowadays: for example, serotonin, which is influential in depression; dopamine, a chemical associated with sensations of pleasure; and epinephrine, more commonly known as adrenaline, a chemical often produced in response to stress, anxiety, and fear, but also critical for attention and vigilance. The scientific term for the transmission of an electrochemical signal from one neuron to another is *action potential*—a term that sounded as strange to me as the word *emptiness* might sound to people who have never been trained as Buddhists.

Recognizing the activity of neurons wouldn't be very important in terms of suffering or happiness, except for a couple of important details. When neurons connect, they form a bond very much like old friendships. They get into a habit of passing the same sorts of messages back and forth, the way old friends tend to reinforce each other's judgments about people, events, and experiences. This bonding is the biological basis for many of what we call mental habits, the kind of "knee-jerk" reactions we have to certain types of people, places, and things.

To use a very simple example, if I'd been frightened by a dog at a very young age, a set of neuronal connections would have been formed in my brain that corresponded to the physical sensations of fear, on one hand, and the concept *dogs are scary,* on the other. The next time I saw a dog, the same set of neurons would start chattering at one another again to remind me that *dogs are scary.* And every time that chatter would occur, it would grow louder and more convincing, until it became such an established routine that all I'd have to do was *think* about dogs and my heart would start pounding and I'd begin to sweat.

But suppose someday I visited a friend who had a dog. Initially, I might feel scared hearing it bark when I knocked on the door and when the animal rushed out to sniff me. But after a while the dog would get used to me and come around to sit by my feet or on my lap, and maybe even start to lick me—so happily and lovingly that I'd practically have to push it away.

What's happened in the dog's brain is that a set of neuronal connections associated with my scent and all the sensations that tell it that its owner likes me creates a pattern that is the equivalent of "Hey, this person is *cool!*" In my own brain, meanwhile, a new set of neuronal connections associated with pleasant physical sensations start chatting with one another, and I'd begin to think, *Hey, maybe dogs are nice!* Every time I visited my friend, this new pattern would be reinforced and the old one would be weakened—until finally I wouldn't be so scared of dogs anymore.

In neuroscientific terms, this capacity to replace old neuronal connections with new ones is referred to as *neuronal plasticity.* The Tibetan term for this capacity is *le-su-rung-wa,* which may be roughly translated into English as "pliability." You can use either term and sound very smart. What it boils down to is that on a strictly cellular level, *repeated experience can change the way the brain works.* This is the *why* behind the *how* of the Buddhist teachings that deal with eliminating mental habits conducive to unhappiness.

THREE BRAINS IN ONE

The Buddha's forms are classified as three. . . .

—GAMPOPA, *The Jewel Ornament of Liberation,*
translated by Khenpo Konchog Gyaltsen Rinpoche

By now it should be clear that the brain is not a single object, and that the answer to a question like "What is it that thinks this food tastes good—or not so good?" is not as simple as it seems. Even relatively basic activities such as eating and drinking involve the exchange of thousands of carefully coordinated, split-second electrochemical signals between millions of cells in the brain and throughout the body. There is, however, an additional level of complexity that must be considered before we complete our tour of the brain.

The billions of neurons in the human brain are grouped by function into three different layers, each of which developed over hundreds of thousands of years as the species evolved and acquired increasingly complex mechanisms for survival. The first and oldest of these layers, known as the *brain stem,* is a bulb-shaped group of cells that extends right out of the top of the spinal cord. This layer is also commonly referred to as the *reptilian brain,* owing to its similarity to the entire brain of many species of reptiles. The primary purpose of the reptilian brain is to regulate basic, involuntary functions such as breathing, metabolism, heartbeat, and circulation. It also controls what is often called the fight-or-flight, or "startle," response: an automatic reaction that compels us to interpret any unexpected encounter or event—for example, a loud noise, an unfamiliar scent, something crawling along our arm, or something coiled in a dark corner—as a possible threat. Without conscious direction, adrenaline starts coursing through the body, the heart speeds up, the muscles tense. If the threat is perceived to be greater than our ability to overcome it, we run. If we think we can beat the threat, we fight. It's easy to see how an automatic response of this sort would greatly affect survival.

Most reptiles tend to be more combative than cooperative, and they possess no innate capacity for nurturing their young. After laying her eggs, the female reptile typically abandons the nest. When the young hatch, though they possess the instincts and capacities of their adult counterparts, their bodies are still vulnerable and awkward, and they must fend for themselves. Many do not survive their first few hours of life. As they scramble toward the safety of whatever habitat is most natural to them—such as the sea, in the case of sea turtles—they're killed and eaten by other animals, and quite often by members of their own species. In fact, it's not uncommon for newly hatched reptiles to be killed by their parents, who don't recognize their prey as their own offspring.

With the evolution of new classes of vertebrates, such as birds and mammals, a startling development in the structure of the brain occurred. Unlike their reptilian cousins, newborn members of these classes aren't sufficiently developed to care for themselves; they require some degree of parental nurturing. To fulfill this need—and to ensure the survival of the species—a second layer of the brain gradually evolved. This layer, referred to as the *limbic region,* surrounds the brain stem like a kind of helmet, and includes a series of programmed neural connections that stimulate the impulse to nurture—that is, to provide food and protection, and to teach essential survival skills through play and other exercises.

These more sophisticated neural pathways also provided new classes of animals with the capacity to distinguish a wider range of emotions than simple fight-or-flight. For example, mammalian parents can distinguish not only the specific sounds made by their own young, but can also differentiate among the types of sounds they make—such as distress, pleasure, hunger, and so on. In addition, the limbic region provides a broader and more subtle capacity to "read" intentions of other animals through physical posture, style of movement, facial expression, the set of the eyes, and even subtle scents or pheromones. And through being able to process these various kinds of signals, mammals and birds are able to adapt more flexibly to changing circumstances, laying the groundwork for learning and memory.

The limbic system has some remarkable structures and capabilities that we'll examine more closely later on, when we look at the role of emotions. Two of its structures, however, deserve special mention. The first is the *hippocampus,* located in the temporal lobe—that is, just behind the temple. (Actually, we have two hippocampi, one on either side of the brain.) The hippocampus is crucial for creating new memories of directly experienced events, providing a spatial, intellectual, and—in the case of human beings at least—verbal context that gives meaning to our emotional responses. People who have suffered physical damage to this region of the brain have difficulty creating new memories; they can remember everything up to the moment the hippocampus was injured, but after the injury they forget, within moments, anyone they meet and anything that happens. The hippocampus is also one of the first areas of the brain to be affected by Alzheimer's disease, as well as by mental illnesses such as schizophrenia, severe depression, and bipolar disorder.

The other significant part of the limbic system is the *amygdala,* a small, almond-shaped neuronal structure situated at the bottom of the limbic region, just above the brain stem. Just as with the hippocampus, there are two of these little organs in the human brain: one in the right hemisphere, the other in the left. The amygdala plays a critical role in both the ability to feel emotions and to create emotional memories. Research has shown that damage to or removal of the amygdala results in a loss of the capacity for almost all types of emotional response, including the most basic impulses of fear and empathy, as well as in an inability to form or to recognize social relationships.[2]

The activity of the amygdala and hippocampus bears close attention as we attempt to define a practical science of happiness. Because the amygdala is connected to the *autonomic nervous system,* the area of the brain stem that automatically regulates muscle, cardiac, and glandular responses, and the *hypothalamus,* a neuronal structure at the base of the limbic region that releases adrenaline and other hormones into the bloodstream, the emotional memories it creates are extremely powerful, linked to significant biological and biochemical reactions.

When an event that generates a strong biological response—such as the release of a large amount of adrenaline or other hormones—occurs, the hippocampus sends a signal down to the brain stem, where it is stored as a pattern. As a result, many people are able to recall exactly where they were and what was going on around them when, for example, they heard about or saw the space shuttle disasters or the assassination of President Kennedy. The same types of patterns can be stored for more personal experiences of a highly charged positive or negative nature.

Because such memories and their associated patterns are so powerful, they can be triggered quite easily by later events that bear some resemblance—sometimes very slight—to the original memory. This type of strong memory response obviously offers important survival benefits in the face of life-threatening circumstances. It allows us to recognize and avoid foods that once made us sick, or avoid confronting especially aggressive animals or members of our own species. But it can also cloud or distort perceptions of more ordinary experiences. For example, children who were regularly humiliated and criticized by their parents or other adults may experience inappropriately strong feelings of fear, resentment, or other unpleasant emotions when dealing with authority figures in adult life. These types of distorted reactions often result from the loose method of association on which the amygdala relies to trigger a memory response. One significant element in a present situation that is similar to an element of a past experience can stimulate the whole range of thoughts, emotions, and hormonal and muscular responses stored with the original experience.

The activities of the limbic system—or the "emotional brain," as it's sometimes referred to—are to a large extent balanced by the third and most recently developed layer of the brain: the *neocortex*. This layer of the brain, which is specific to mammals, provides the capacity for reasoning, forming concepts, planning, and fine-tuning emotional responses. Though it's fairly thin in most mammals, anyone who has ever witnessed a cat figure out how to pry open a closet door or watched a dog learn to manipulate a door handle has seen an animal's neocortex at work.

Among humans and other highly evolved mammals the neocortex developed into a much larger and more complicated structure. When most of us think of the brain, it's usually this structure—with its many bulges and grooves—that appears in our mind's eye. In fact, were it not for these bulges and grooves, we wouldn't be able to imagine the brain at all, since our large neocortex provides us with the capacity for imagination, as well as the ability to create, understand, and manipulate symbols. It's our neocortex that provides us with our capacity for language, writing, mathematics, music, and art. Our neocortex is the seat of our rational activities, including problem solving, analysis, judgment, impulse control, and the abilities to organize information, learn from past experiences and mistakes, and empathize with others.

Simply recognizing that the human brain is composed of these three different layers is itself amazing. Even more fascinating, however, is that no matter how modern or sophisticated we think we are, the production of a single thought requires a series of complex interactions among all three layers of the brain—the brain stem, the limbic region, and the neocortex. In addition, it appears that *every* thought, sensation, or experience involves a *different* set of interactions, often involving areas of the brain that aren't activated by other types of thoughts.

THE MISSING CONDUCTOR

The mind is not in the head.

> —Francisco J. Varela, "Steps to a Science of Inter-Being,"
> from *The Psychology of Awakening,*
> edited by S. Bachelor, G. Claxton, and G. Watson

One question still troubled me, though. If the brain is a symphony, as Dr. Livingston suggested, shouldn't there be a conductor? Shouldn't there be some objectively identifiable cell or organ that directs everything? We certainly feel as though there is such a thing—or at least we refer to it when we say things like "I haven't made up my mind," or "My mind's a total blank," or "I must have been out of my mind."

From what I've learned in talking with neuroscientists, biologists, and psychologists, modern science has been looking for such a "conductor" for a long time, investing a great deal of effort in hopes of discovering some cell or group of cells that directs sensation, perception, thought, and other kinds of mental activity. Yet, so far, even using the most sophisticated technology available, no evidence of a conductor has been found. There's no single area—no tiny "self"—in the brain that can be said to be responsible for coordinating the communication among the different players.

Contemporary neuroscientists have thus abandoned the search for a "conductor" in favor of exploring the principles and mechanisms by which billions of neurons distributed across the brain are able to coordinate their activity harmoniously, without the need of a central director. This "global" or "distributed" behavior may be compared to the spontaneous coordination of a group of jazz musicians. When jazz musicians are improvising, each of them may be playing a slightly different musical phrase, yet somehow they still manage to play together harmoniously.

The idea of locating a "self" in the brain was based, in many ways, on the influence of classical physics, which had traditionally focused on studying the laws governing localized objects. Based on this traditional viewpoint, if the mind had an effect—for instance, on emotions—it should be localized somewhere. Yet the whole idea of solid entities is questionable within the framework of contemporary physics. Every time someone identifies the tiniest element of matter imaginable, someone else discovers that it's actually made up of even tinier particles. With each new advance, it's becoming more difficult to conclusively identify any fundamental material element.

Logically speaking, then, even if it were possible to dissect the brain into smaller and smaller pieces, down to the smallest subatomic level, how could anyone be sure of precisely identifying a single one of these pieces as the mind? Since every cell is made up of many smaller particles, each of which is made up of even smaller particles, how would it be possible to recognize which one constitutes the mind?

It's on this point that Buddhism may be able to offer a fresh perspective, one that can perhaps form the basis for new avenues of scientific research. The Tibetan Buddhist term for mind is *sem,* a word that may be translated into English as "that which knows." This simple term can help us to understand the Buddhist view of the mind as less of a specific object than of a *capacity* to recognize and reflect on our experiences. Although the Buddha taught that the brain is, indeed, the physical support for the mind, he was also careful to point out that *the mind itself* isn't something that can be seen, touched, or even defined by words. Just as the physical organ of the eye is not sight, and the physical organ of the ear is not hearing, the brain is not the mind.

One of the earliest lessons I was taught by my father was that Buddhists don't see the mind as a discrete entity, but rather as a perpetually unfolding experience. I can remember how strange this idea seemed to me at first, sitting in the teaching room of his monastery in Nepal, surrounded by students from around the world. There were so many of us crammed together in this tiny room that there was barely enough space to move. But from the windows I could see a huge expanse of mountains and forests. And my father was sitting there, very composed, oblivious of the heat generated by so many people, saying that what we think of as our identity—"my mind," "my body," "my self"—is actually an illusion generated by the unceasing flow of thoughts, emotions, sensations, and perceptions.

I don't know whether it was the sheer force of my father's own experience as he spoke, or the physical contrast between feeling crammed on a bench among other students and the view through the window of the open spaces beyond, or both—but in that moment something, as they say in the West, simply "clicked." I had an experience of the freedom of distinguishing between thinking in terms of "my" mind or "my" self and the possibility of simply experiencing *being* as widely and openly as the expanse of mountains and sky beyond the windows.

Later, when I came to the West, I heard a number of psychologists compare the experience of "mind" or "self" to watching a movie. When we watch a movie, they explained, we seem to experience a continuous flow of sound and motion as individual frames pass through a projector. The experience would be drastically different, however, if we had the chance to look at the film frame by frame.

This is exactly how my father began to teach me to look at my mind. If I observed every thought, feeling, and sensation that passed through my mind, the illusion of a limited self would dissolve, to be replaced by a sense of awareness that is much more calm, spacious, and serene. And what I learned from other scientists was that because experience changes the neuronal structure of the brain, when we observe the mind this way, we can change the cellular gossip that perpetuates our experience of our "self."

MINDFULNESS

Looking again and again at the mind which cannot be looked at,
the meaning can be vividly seen, just as it is.
—THE THIRD GYALWANG KARMAPA,
  Song of Karmapa: The Aspiration of the Mahamudra of True Meaning,
  translated by Erik Pema Kunsang

The key—the *how* of Buddhist practice—lies in learning to simply rest in a bare awareness of thoughts, feelings, and perceptions as they occur. In the Buddhist tradition, this gentle awareness is known as *mindfulness,* which, in turn, is simply resting in the mind's natural clarity. Just as in the example of the dog, if I were to become aware of my habitual thoughts, perceptions, and sensations, rather than being carried away by them, their power over me would begin to fade. I would experience their coming and going as nothing more than the natural function of the mind, in the same way that waves naturally ripple across the surface of a lake or ocean. And ultimately, I realize, this is exactly what happened when I sat alone in my retreat room trying to overcome the anxiety that had made me so uncomfortable throughout my childhood. Simply *looking* at what was going on in my mind actually changed what was going on there.

You can begin to taste the same freedom of natural clarity right now, through a simple exercise. Simply sit up straight, breathe normally, and allow yourself to become aware of your breath coming in and going out. As you relax into simply being aware of your inhalation and exhalation, you'll probably start to notice hundreds of thoughts passing through your mind. Some of them are easy to let go of, while others may lead you down a long avenue of related thoughts. When you find yourself chasing after a thought, simply bring yourself back to focusing on your breath. Do this for about a minute.

In the beginning, you may be surprised by the sheer number and variety of thoughts that pour through your awareness like a waterfall rushing over a steep cliff. An experience of this sort is not a sign of failure. It's a sign of success. You've begun to recognize how many thoughts ordinarily pass through your mind without your even noticing them.

You may also find yourself getting caught up in a particular train of thought and following it while ignoring everything else. Then suddenly you remember that the point of the exercise is simply to watch your thoughts. Instead of punishing or condemning yourself, just go back to focusing on your breath.

If you keep up this practice, you'll find that even though thoughts and emotions come and go, the mind's natural clarity is never disturbed or interrupted. To use an example, during a trip to Nova Scotia, I visited a retreat house that was quite near the ocean. The day I arrived, the weather was perfect: The sky was cloudless and the ocean was a deep, clear blue—very pleasant to look at. When I woke the next morning, though, the ocean looked like a thick, muddy soup. I wondered, "What happened to the ocean? Yesterday it was so clear and blue, and today it's suddenly dirty." I took a walk down to the shore, and couldn't see any obvious reason for the change. There wasn't any mud in the water or along the beach. Then I looked up at the sky and saw that it was thick with dark greenish clouds; and I realized it was the color of the clouds that had changed the color of the ocean. The water itself, when I looked closely at it, was still clean and clear.

The mind, in many ways, is like the ocean. The "color" changes from day to day or moment to moment, reflecting the thoughts, emotions, and so on passing "overhead," so to speak. But the mind itself, like the ocean, never changes: It's always clean and clear, no matter what it's reflecting.

Practicing mindfulness may seem hard at first, but the point is not how successful you are right away. What seems impossible at present becomes easier with practice. There's nothing you can't get used to. Just think about all the unpleasant things you've accepted as ordinary, like wading through traffic or dealing with a bad-tempered relative or coworker. Becoming mindful is a gradual process of establishing new neuronal connections and inhibiting the gossip among old ones. It requires patiently taking one small step at a time, practicing in very short intervals.

There's an old Tibetan saying: "If you walk with haste, you won't reach Lhasa. Walk gently and you'll reach your goal." This proverb comes from the days when people in eastern Tibet would make a pilgrimage to Lhasa, the capital city, in the central region of the country. Pilgrims who wanted to get there quickly would walk as fast as they could, but because of the pace they set for themselves, they'd get tired or sick and have to return home. Those who traveled at an easy pace, however, pitched camp for the night, enjoyed one another's company, and then continued on the next day, actually arrived at Lhasa more quickly.

Experience follows intention. Wherever we are, whatever we do, all we need to do is recognize our thoughts, feelings, and perceptions as something natural. Neither rejecting nor accepting, we simply acknowledge the experience and let it pass. If we keep this up, we'll eventually find ourselves becoming able to manage situations we once found painful, scary, or sad. We'll discover a sense of confidence that isn't rooted in arrogance or pride. We'll realize that we're always sheltered, always safe, and always home.

Remember that little test I asked you to try, about asking yourself the next time you sit down to lunch or dinner, "What is it that thinks that this food tastes good—or not so good? What is it that recognizes eating?" It seemed pretty easy to answer, once upon a time. But the answer doesn't come so easily anymore, does it?

Even so, I'd like you to try it again the next time you sit down to lunch or dinner. If the answers that come up for you now are confusing and conflicting, that's good. Confusion, I was taught, is the beginning of understanding, the first stage of letting go of the neuronal gossip that used to keep you chained to very specific ideas about who you are and what you're capable of.

Confusion, in other words, is the first step on the path to real well-being.

3

BEYOND THE MIND, BEYOND THE BRAIN

When the mind is realized, that is the buddha.
                    —*The Wisdom of the Passing Moment Sutra,*
                    translated by Elizabeth M. Callahan

YOU'RE NOT THE limited, anxious person you think you are. Any trained Buddhist teacher can tell you with all the conviction of personal experience that, really, you're the very heart of compassion, completely aware, and fully capable of achieving the greatest good, not only for yourself, but for everyone and everything you can imagine.

The only problem is that you don't recognize these things about yourself. In the strictly scientific terms I've come to understand through conversations with specialists in Europe and North America, most people simply mistake the habitually formed, neuronally constructed image of themselves for who and what they really are. And this image is almost always expressed in dualistic terms: self and other, pain and pleasure, having and not having, attraction and repulsion. As I've been given to understand, these are the most basic terms of survival.

Unfortunately, when the mind is colored by this dualistic perspective, every experience—even moments of joy and happiness—is bounded by some sense of limitation. There's always a *but* lurking in the background. One kind of *but* is the *but* of difference. "Oh, my birthday party was wonderful, but I would have liked chocolate cake instead of carrot cake." Then there's the *but* of "better." "I love my new house, but my friend John's place is bigger and has much better light."

And finally, there's the *but* of fear. "I can't stand my job, but in this market how will I ever find another one?" Personal experience has taught me that it's possible to overcome any sense of personal limitation. Otherwise I'd probably still be sitting in my retreat room, feeling too scared and inadequate to participate in group practices. As a thirteen-year-old boy, I only knew *how* to get over my fear and insecurity. Through the patient tutoring of experts in the fields of psychology and neuroscience, like Francisco Varela, Richard Davidson, Dan Goleman, and Tara Bennett-Goleman, I've begun to recognize *why*, from an objectively scientific perspective, the practices actually work: that feelings of limitation, anxiety, fear, and so on are just so much neuronal gossip. They are, in essence, habits. And habits can be unlearned.

NATURAL MIND

> It is called "true nature" because no one created it.
>
> —CHANDRAKIRTI, *Entering the Middle Way,*
> translated by Ari Goldfield

One of the first things I learned as a Buddhist was that the fundamental nature of the mind is so vast that it completely transcends intellectual understanding. It can't be described in words or reduced to tidy concepts. For someone like me, who likes words and feels very comfortable with conceptual explanations, this was a problem.

In Sanskrit, the language in which the Buddha's teachings were originally recorded, the fundamental nature of the mind is called *tathagatagarbha*, which is a very subtle and tricky description. Literally, it means "the nature of those who have gone that way." "Those who have gone that way" are the people who have attained complete enlightenment—in other words, people whose minds have completely surpassed ordinary limitations that can be described in words.

Not a lot of help there, I think you'll agree.

Other, less literal translations have variously rendered tathagatagarbha as "Buddha nature," "true nature," "enlightened essence," "ordinary mind," and even "natural mind"—none of which sheds much light on the real meaning of the word itself. To really understand tathagata-garbha, you have to experience it directly, which for most of us occurs initially in the form of quick, spontaneous glimpses. And when I finally experienced my first glimpse, I realized that everything the Buddhist texts said about it was true.

For most of us, our natural mind or Buddha nature is obscured by the limited self-image created by habitual neuronal patterns—which, in themselves, are simply a reflection of the unlimited capacity of the mind to create any condition it chooses. Natural mind is capable of producing anything, *even ignorance of its own nature.* In other words, not recognizing natural mind is simply an example of the mind's unlimited capacity to create whatever it wants. Whenever we feel fear, sadness, jealousy, desire, or any other emotion that contributes to our sense of vulnerability or weakness, we should give ourselves a nice pat on the back. We've just experienced the unlimited nature of the mind.

Although the true nature of the mind can't be described directly, that doesn't mean we shouldn't at least try to develop some theoretical understanding about it. Even a limited understanding is at least a signpost, pointing the way toward direct experience. The Buddha understood that experiences impossible to describe in words could best be explained through stories and metaphors. In one text, he compared tathagatagarbha to a nugget of gold covered with mud and dirt.

Imagine you're a treasure hunter. One day you discover a chunk of metal in the ground. You dig a hole, pull out the metal, take it home, and start to clean it. At first, one corner of the nugget reveals itself, bright and shining. Gradually, as you wash away the accumulated dirt and mud, the whole chunk is revealed as gold. So let me ask: Which is more valuable—the chunk of gold buried in mud, or the one you cleaned? Actually, the value is equal. Any difference between the dirty nugget and the clean is superficial.

The same can be said of natural mind. The neuronal gossip that keeps you from seeing your mind in its fullness doesn't really change the fundamental nature of your mind. Thoughts like "I'm ugly," "I'm stupid," or "I'm boring" are nothing more than a kind of biological mud, temporarily obscuring the brilliant qualities of Buddha nature, or natural mind.

Sometimes the Buddha compared natural mind to space, not necessarily as space is understood by modern science, but rather in the poetic sense of the profound experience of openness one feels when looking up at a cloudless sky or entering a very large room. Like space, natural mind isn't dependent on prior causes or conditions. It simply *is*: immeasurable and beyond characterization, the essential background through which we move and relative to which we recognize distinctions between the objects we perceive.

NATURAL PEACE

> *In natural mind, there is no rejection or acceptance, no loss or gain.*
>
> —THE THIRD GYALWANG KARMAPA,
>   *Song of Karmapa: The Aspiration of the Mahamudra of True Meaning*,
>   translated by Erik Pema Kunsang

I'd like to make it clear that the comparison between natural mind and space as described by modern science is really more of a useful metaphor than an exact description. When most of us think of space, we think of a blank background against which all sorts of things appear and disappear: stars, planets, comets, meteors, black holes, and asteroids—even things that haven't yet been discovered. Yet, despite all this activity, our idea of the essential nature of space remains undisturbed. As far as we know, at least, space has yet to complain about what happens within itself. We've sent thousands—millions—of messages out into the universe, and never once have we received a response like "I am so angry that an asteroid just smashed into my favorite planet" or "Wow, I'm thrilled! A new star has just come into being!"

In the same way, the essence of mind is untouched by unpleasant thoughts or conditions that might ordinarily be considered painful. It's naturally peaceful, like the mind of a young child accompanying his parents through a museum. While his parents are completely caught up in judging and evaluating the various works of art on display, the child merely sees. He doesn't wonder how much some piece of art might have cost, how old a statue is, or whether one painter's work is better than another's. His perspective is completely innocent, accepting everything it beholds. This innocent perspective is known in Buddhist terms as "natural peace," a condition similar to the sensation of total relaxation a person experiences after, say, going to the gym or completing a complicated task.

This experience is illustrated very nicely by an old story about a king who had ordered the construction of a new palace. When the new building was finished, he was faced with the problem of secretly transferring all his treasure—gold, jewels, statues, and other objects—from the old palace to the new one. He couldn't perform this task by himself, because his time was taken up with performing all his royal duties, but there weren't many people in his court that he could trust to carry out the job without stealing some of the treasure for themselves. He knew of one loyal general, though, whom he could trust to carry out the job in complete secrecy and with great efficiency.

So the king summoned the general and explained to him that, as he was the only trustworthy person at court, he would like him to take upon himself the task of moving all the treasures from the old palace to the new one. The most important part of the job, aside from the secrecy, was that the transfer had to be completed in a single day. If the general could accomplish this, the king promised in return to bestow upon him vast tracts of rich farmland, stately mansions, gold, jewels—enough wealth, in fact, to allow him to retire in comfort for the rest of his life. The general accepted the assignment willingly, dazzled by the prospect of being able to accumulate enough wealth in a single day's work to guarantee that his children, grandchildren, and great-grandchildren could spend their days in comfort and splendor.

The general woke early the next morning and set himself to moving the king's treasures from the old palace to the new one, running back and forth along secret passageways with boxes and chests of gold and jewels, and allowing himself only one brief rest for lunch to keep his strength up. Fnally he succeeded in transferring the last of the king's treasure to the storehouse in the new palace, and just as the sun set, he went to the king and reported that the task was complete. The king congratulated him and handed him all the deeds and titles to the rich lands he'd been promised, and the gold and jewels that were part of the bargain as well.

When he returned home, the general took a hot bath, dressed himself in comfortable robes, and, with a deep sigh, settled himself on a pile of soft cushions in his private room, exhausted but contented that he'd successfully completed the incredibly difficult task he'd been assigned. Experiencing a complete sense of confidence and accomplishment, he was just able to let go and experience the freedom of being exactly as he was in that moment.

This perfectly effortless state of relaxation is what is meant by natural peace.

As with so many aspects of natural mind, the experience of natural peace is so far beyond what we normally consider relaxation that it defies description. In classical Buddhist texts, it's compared to offering candy to a mute. The mute undoubtedly experiences the sweetness of the candy, but is powerless to describe it. In the same way, when we taste the natural peace of our own minds, the experience is unquestionably real, yet beyond our capacity to express in words.

So now, the next time you sit down to eat, if you should ask yourself, "What is it that thinks that this food tastes good—or not so good? What is it that recognizes eating?" don't be surprised if you can't answer at all. Congratulate yourself instead. When you can't describe a powerful experience in words anymore, it's a sign of progress. It means you've at least dipped your toes into the realm of the ineffable vastness of your true nature, a very brave step that many people, too comfortable with the familiarity of their discontent, lack the courage to take.

The Tibetan word for meditation, *gom*, literally means "becoming familiar with," and Buddhist meditation practice is really about becoming familiar with the nature of your own mind—a bit like getting to know a friend on deeper and deeper levels. Also like getting to know a friend, discovering the nature of your mind is a gradual process.

Rarely does it occur all at once. The only difference between meditation and ordinary social interaction is that the friend you're gradually
coming to know is yourself.

GETTING TO KNOW YOUR NATURAL MIND

*If an inexhaustible treasure were buried in the ground beneath
a poor man's house, the man would not know of it, and the
treasure would not speak and tell him, "I am here!"*

> —MAITREYA, *The Mahayana Uttaratantra Shastra,*
> translated by Rosemarie Fuchs

The Buddha often compared natural mind to water, which in its
essence is always clear and clean. Mud, sediment, and other impurities may temporarily darken or pollute the water, but we can filter
away such impurities and restore its natural clarity. If water weren't
naturally clear, no matter how many filters you used, it would not become clear.

The first step toward recognizing the qualities of natural mind is illustrated by an old story told by the Buddha, about a very poor man
who lived in a rickety old shack. Though he didn't know it, hundreds
of gems were embedded in the walls and floor of his shack. Though he
owned all those jewels, because he didn't understand their value, he
lived as a pauper—suffering from hunger and thirst, the bitter cold of
winter and the terrible heat of summer.

One day a friend of his asked him, "Why are you living like such a
pauper? You're not poor. You're a very rich man."

"Are you crazy?" the man replied. "How can you say such a thing?"

"Look around you," his friend said. "Your whole house is filled with
jewels—emeralds, diamonds, sapphires, rubies."

At first the man didn't believe what his friend was saying. But after
a while he grew curious, and took a small jewel from his walls into
town to sell. Unbelievably, the merchant to whom he brought it paid
him a very handsome price, and with the money in hand, the man returned to town and bought a new house, taking with him all the jewels

he could find. He bought himself new clothes, filled his kitchen with food, engaged servants, and began to live a very comfortable life.

Now let me ask a question. Who is wealthier—the man who lives in an old house surrounded by jewels he doesn't recognize, or someone who understands the value of what he has and lives in total comfort?

Like the question posed earlier about the nugget of gold, the answer here is: both. They both owned great wealth. The only difference is that for many years one didn't recognize what he possessed. It wasn't until he recognized what he already had that he freed himself from poverty and pain.

It's the same for all of us. As long as we don't recognize our real nature, we suffer. When we recognize our nature, we become free from suffering. Whether you recognize it or not, though, its qualities remain unchanged. But when you begin to recognize it in yourself, you change, and the quality of your life changes as well. Things you never dreamed possible begin to happen.

MIND, BIOLOGY, OR BOTH?

> *The buddha abides in your own body. . . .*
>
> —The Samputa Tantra,
> translated by Elizabeth M. Callahan

Just because something hasn't been identified doesn't mean it doesn't exist. We've already seen this in the attempt to concretely identify the location of the mind: While there's ample evidence of mental activity, no scientist has been able to confirm the existence of the mind itself. Likewise, no scientist has been able to precisely define the nature and properties of space at the most fundamental level. Yet we know we have a mind, and we can't deny the existence of space. Mind and space are concepts deeply ingrained in our culture. We're familiar with these ideas. They feel normal and, to some degree, quite ordinary.

Notions such as "natural mind" and "natural peace" don't enjoy the same degree of familiarity, however. Consequently, many people approach them with a certain amount of skepticism. Yet it would be fair to say that by using the same processes of inference and direct experience, we can gain at least some familiarity with natural mind.

The Buddha taught that the reality of natural mind could be demonstrated by a certain sign obvious to everyone, posed in the form of a question and an answer. The question was "In general, what is the one area of concern shared by all people?"

When I ask this same question in public teachings, people give a number of different answers. Some people answer that the main concern is staying alive, being happy, avoiding suffering, or being loved. Other replies include peace, progress, eating, and breathing; not changing anything; and improving living circumstances. Still other responses include being in harmony with oneself and others or understanding the meaning of life or the fear of death. One answer I find especially funny is "Me!"

Every answer is absolutely correct. They just represent different aspects of the ultimate reply.

The basic concern shared by all beings—humans, animals, and insects alike—is the desire to be happy and to avoid suffering.

Although each of us may have a different strategy, in the end we're all working for the same result. Even ants never stay still, even for a second. They're running around all the time, gathering food and building or expanding their nests. Why do they go to so much trouble? To find some kind of happiness and avoid suffering.

The Buddha said that the desire to achieve lasting happiness and to avoid unhappiness is the one unmistakable sign of the presence of natural mind. There are in fact many other indicators, but listing them all would probably require another book. So why did the Buddha assign such importance to this one particular sign?

Because the true nature of all living creatures is *already* completely free from suffering and endowed with perfect happiness: In seeking happiness and avoiding unhappiness, regardless of how we go about it, we're all just expressing the essence of who we are.

The yearning most of us feel for a lasting happiness is the "small, still voice" of the natural mind, reminding us of what we're really capable of experiencing. The Buddha illustrated this longing through the example of a mother bird that has left her nest. No matter how beautiful the place she has flown to, no matter how many new and interesting things she sees there, something keeps pulling her to return to her nest. In the same way, no matter how absorbing daily life might be—no matter how great it may temporarily feel to fall in love, receive praise, or get the "perfect" job—the yearning for a state of complete, uninterrupted happiness pulls at us.

In a sense, we're homesick for our true nature.

BEING YOU

We need to recognize our basic state.

—TSOKNYI RINPOCHE, *Carefree Dignity,*
translated by Erik Pema Kunsang and Marcia Binder Schmidt

According to the Buddha, the basic nature of mind can be directly experienced simply by allowing the mind to rest simply as it is. How do we accomplish this? Let's return to the story about the general charged with moving the king's treasure from one place to another in a single day, and remember how relaxed and contented he felt once he'd finished the job. As he sat on his cushions after his bath, his mind was completely at rest. Thoughts were still bubbling up, but he was content to let them rise and fall without hanging on to any of them or following any of them through.

You've probably experienced something similar after finishing a long and difficult job, whether it involved physical labor or the type of mental effort involved in writing a report or completing some sort of financial analysis. When you finish the job, your mind and body naturally come to rest in a state of happy exhaustion.

So let's try a brief exercise in resting the mind. This is not a meditation exercise. In fact, it's an exercise in "non-meditation"—a very old Buddhist practice that, as my father explained it, takes the pressure off thinking you have to achieve a goal or experience some sort of special state. In non-meditation, we just watch whatever happens without interfering. We're merely interested observers of a kind of introspective experiment, with no investment in how the experiment turns out.

Of course, when I first learned this, I was still a pretty goal-oriented child. I wanted something wonderful to happen every time I sat down to meditate. So it took me a while to get the hang of just resting, just looking, and letting go of the results.

First, assume a position in which your spine is straight, and your body is relaxed. Once your body is positioned comfortably, allow your mind to simply rest for three minutes or so. Just let your mind go, as though you've just finished a long and difficult task.

Whatever happens, whether thoughts or emotions occur, whether you notice some physical discomfort, whether you're aware of sounds or smells around you, or your mind is a total blank, don't worry. Anything that happens—or doesn't happen—is simply part of the experience of allowing your mind to rest.

So now, just rest in the awareness of whatever is passing through your mind. . . .

Just rest. . . .

Just rest. . . .

When the three minutes are up, ask yourself, How was that experience? Don't judge it; don't try to explain it. Just review what happened and how you felt. You might have experienced a brief taste of peace or openness. That's good. Or you might have been aware of a million different thoughts, feelings, and sensations. That's also good. Why? Because either way, as long as you've maintained at least a bare awareness of what you were thinking or feeling, you've had a direct glimpse of your own mind just performing its natural functions.

So let me confide in you a big secret. Whatever you experience when you simply rest your attention on whatever's going on in your mind at any given moment *is* meditation. Simply resting in this way *is* the experience of natural mind.

The only difference between meditation and the ordinary, everyday process of thinking, feeling, and sensation is the application of the simple, bare awareness that occurs when you allow your mind to rest simply as it is—without chasing after thoughts or becoming distracted by feelings or sensations.

It took me a long time to recognize how easy meditation really is, mainly because it seemed so completely ordinary, so close to my everyday habits of perception, that I rarely stopped to acknowledge it. Like many of the people I now meet on teaching tours, I thought that natural mind had to be something else, something different from, or better than, what I was already experiencing.

Like most people, I brought so much judgment to my experience. I believed that thoughts of anger, anxiety, fear, and so on that came and went throughout the day were bad or counterproductive—or at the very least inconsistent with natural peace! The teachings of the Buddha—and the lesson inherent in this exercise in non-meditation—is that if we allow ourselves to relax and take a mental step back, we can begin to recognize that all these different thoughts are simply coming and going within the context of an unlimited mind, which, like space, remains fundamentally unperturbed by whatever occurs within it.

In fact, experiencing natural peace is easier than drinking water. In order to drink, you have to expend effort. You have to reach for the glass, bring it to your lips, tip the glass so the water pours into your mouth, swallow the water, and then put the glass down. No such effort is required to experience natural peace. All you have to do is rest your mind in its natural openness. No special focus, no special effort, is required.

And if for some reason you cannot rest your mind, you can simply observe whatever thoughts, feelings, or sensations come up, hang out for a couple of seconds, and then disappear, and acknowledge, "Oh, that's what's going on in my mind right now."

Wherever you are, whatever you do, it's essential to acknowledge your experience as something ordinary, the natural expression of your true mind. If you don't try to stop whatever is going on in your mind, but merely observe it, eventually you'll begin to feel a tremendous sense of relaxation, a vast sense of openness within your mind—which is in fact your natural mind, the naturally unperturbed *background* against which various thoughts come and go. At the same time, you'll be awakening new neuronal pathways, which, as they grow stronger and more deeply connected, enhance your capacity to tolerate the cascade of thoughts rushing through your mind at any given moment. Whatever disturbing thoughts do arise will act as catalysts that stimulate your awareness of the natural peace that surrounds and permeates these thoughts, the way space surrounds and permeates every particle of the phenomenal world.

But now it's time to leave the general introduction to the mind and begin to examine its characteristics in more detail. You may wonder why it's necessary to know anything more about natural mind. Isn't a general understanding enough? Can't we just skip to the practices right now?

Think of it this way: If you were driving in the dark, wouldn't you feel better having a map of the terrain, instead of just a rough idea of where you were going? Without a map, and without any signs to guide you, you could get lost. You might find yourself taking all sorts of wrong turns and side roads, making the trip longer and more complicated than necessary. You could wind up traveling in circles. Sure, you might eventually end up where you want to go—but the journey would be a lot easier if you knew where you were going. So think of the next two chapters as a map, a set of guidelines and signposts that can help you get where you want to go more quickly.

4

EMPTINESS: THE REALITY
BEYOND REALITY

Emptiness is described as the basis that makes everything possible.

— THE TWELFTH TAI SITUPA RINPOCHE,
Awakening the Sleeping Buddha

THE SENSE OF openness people experience when they simply rest their minds is known in Buddhist terms as *emptiness*, which is probably one of the most misunderstood words in Buddhist philosophy. It's hard enough for Buddhists to understand the term, but Western readers have an even more difficult time, because many of the early translators of Sanskrit and Tibetan Buddhist texts interpreted *emptiness* as "the Void" or nothingness—mistakenly equating emptiness with the idea that nothing at all exists. Nothing could be further from the truth the Buddha sought to describe.

While the Buddha did teach that the nature of the mind—in fact, the nature of all phenomena—is *emptiness*, he didn't mean that their nature was truly *empty*, like a vacuum. He said it was empti*ness*, which in the Tibetan language is made up of two words: *tongpa-nyi*. The word *tongpa* means "empty," but only in the sense of something beyond our ability to perceive with our senses and our capacity to conceptualize. Maybe a better translation would be "inconceivable" or "unnamable." The word *nyi*, meanwhile, doesn't have any particular meaning in everyday Tibetan conversation. But when added to another word it conveys a sense of "possibility"—a sense that anything can arise, anything can happen. So when Buddhists talk about *emptiness*, we don't mean nothingness, but rather an unlimited potential for anything to appear, change, or disappear.

Perhaps we can use an analogy here to what contemporary physicists have learned about the strange and wonderful phenomena they see when they examine the inner workings of an atom. According to the physicists with whom I've spoken, the basis from which all subatomic phenomena arise is often referred to as the *vacuum state,* the state of lowest energy in the subatomic universe. In the vacuum state, particles continually appear and disappear. So, although seemingly empty, this state is actually very active, full of the potential to produce anything whatsoever. In this sense, the vacuum shares certain qualities with the "empty quality of the mind." Just as the vacuum is considered "empty," yet is the source from which all manner of particles appear, the mind is essentially "empty" in that it defies absolute description. Yet out of this indefinable and incompletely knowable basis, all thoughts, emotions, and sensations perpetually arise.

Because the nature of your mind is emptiness, you possess the capacity to experience a potentially unlimited variety of thoughts, emotions, and sensations. Even misunderstandings of emptiness are simply phenomena arising out of emptiness!

A simple example may help you gain some understanding of emptiness on an experiential level.

A few years ago, a student came to me asking for a teaching on emptiness. I gave him the basic explanation and he appeared to be quite happy—thrilled, in fact.

"That's so cool!" he replied at the end of our conversation.

My own experience had taught me that emptiness isn't so easy to understand after one lesson, so I instructed him to spend the next several days meditating on what he'd learned.

A few days later the student suddenly arrived outside my room with an expression of terror on his face. Pale, hunched, and shaking, he stepped carefully across the room, like someone testing the ground in front of him for quicksand.

When he finally stopped in front of where I was sitting, he said, "Rinpoche, you told me to meditate on emptiness. But last night, it occurred to me that if everything is emptiness, then this whole building is emptiness, the floors are emptiness, and the ground underneath is emptiness. If that's the case, why shouldn't we all fall through the floor and down through the ground?"

I waited until he finished speaking. Then I asked, "Who would fall?"

He thought about the question for a moment, and then his expression changed completely. "Oh," he exclaimed, "I get it! If the building is emptiness and people are emptiness, there's no one to fall and nothing to fall through."

He gave a long sigh, his body relaxed, and the color returned to his face. So I asked him again to meditate on emptiness with this new understanding.

Two or three days later he again arrived at my room unexpectedly. Pale and shaking again, he entered the room, and it seemed quite clear he was trying as best he could to hold his breath, terrified of exhaling.

Sitting down in front of me, he said, "Rinpoche, I meditated on emptiness as you instructed, and I understood that just like this building and the ground below are emptiness, I'm also emptiness. But as I kept pursuing this meditation, I kept going deeper and deeper, until I stopped being able to see or feel anything. I'm so afraid that if I'm nothing more than emptiness, I'm just going to die. That's why I ran to see you this morning. If I'm just emptiness, then I'm basically nothing, and there's nothing to keep me from just dissolving away into nothingness."

When I was sure he was finished, I asked, "Who is it that would dissolve?"

I waited a few moments for him to absorb this question, then pressed on. "You've mistaken emptiness for nothingness. Almost everybody makes the same mistake in the beginning, trying to understand emptiness as an idea or a concept. I made the same mistake myself. But there's really no way to understand emptiness conceptually. You can only really recognize it through direct experience. I'm not asking you to believe me. All I'm saying is that the next few times you sit down to meditate, ask yourself, 'If the nature of everything is emptiness, who or what can dissolve? Who or what is born and who or what can die?' Try that, and the answer you get may surprise you."

After a sigh, he agreed to try again.

Several days later he returned to my room, smiling peacefully as he announced, "I think I'm starting to I understand emptiness."

I asked him to explain.

"I followed your instructions, and after meditating on the subject for a long time, I realized that emptiness isn't nothingness, bacause there must be *something* before there can be *nothing*. Emptiness is everything—all possibilities of existence and nonexistence imaginable, occurring simultaneously. So if our true nature is emptiness, then nobody can be said to truly die and no one can be said to be truly born, because the possibility of being in a certain way and not being in a certain way is present within us at every moment."

"Very good," I told him. "Now forget everything you just said, because if you try to remember it exactly, you'll turn everything you learned into a concept, and we'll have to start all over again."

TWO REALITIES: ABSOLUTE AND RELATIVE

> *Ultimate truth cannot be taught without basis on*
> *relative truth. . . .*
>
> —NĀGARUNĀ, *Mādhyamakārikā,*
> translated by Maria Montenegro

Most of us require time to contemplate and meditate in order to comprehend emptiness. When I teach on this subject, one of the first questions I'm usually asked is "Well, if the basis of reality is emptiness, where does everything come from?" It's a good question, in fact a very profound one. But the relationship between emptiness and experience isn't so simple—or rather, it's so simple that it's easy to miss. It's actually out of the unlimited potential of emptiness that phenomena—a catch-all term that includes thoughts, emotions, sensations, and even material objects—can appear, move, change, and ultimately vanish.

Instead of going into a discussion of quantum mechanics—the contemporary branch of physics that examines matter on atomic and subatomic levels—which I admit is not my area of expertise, I've found that the best way to describe this aspect of emptiness is by going back to the analogy of space as understood in the Buddha's time—a vast openness that is not a thing in itself, but rather an infinite, uncharacterized background against and through which galaxies, stars, planets, animals, human beings, rivers, trees, and so forth appear and move. In the absence of space, none of these things could appear distinct or individual. There would be no room for them, no background against which they could be seen. Stars and planets can only come into being, move about, and dissolve against the background of space. We ourselves are able to stand, sit, and walk in and out of a room only because of the space that surrounds us. Our own bodies are filled with space: the external openings that allow us to breathe, swallow, speak, and so on, as well as the space within our internal organs, such as the lungs that open and close as we inhale and exhale.

A similar relationship exists between emptiness and phenomena. Without emptiness, nothing could appear; in the absence of phenomena, we wouldn't be able to experience the background of emptiness out of which everything appears. So, in a sense, you'd have to say that there is a relationship between emptiness and phenomena. But there's also an important distinction. Emptiness, or infinite possibility, is the *absolute* nature of reality. Everything that appears out of emptiness— stars, galaxies, people, tables, lamps, clocks, and even our perception of time and space—is a *relative* expression of infinite possibility, a momentary appearance in the context of infinite time and space.

I'd like to take this moment to point out another, extremely important, distinction between absolute and relative reality. According to Buddhist understanding and also, apparently, to some modern Western schools of scientific thought, only something that doesn't change, that can't be affected by time and circumstance, or broken down into smaller, connected parts, can be said to be absolutely real. Using this definition as a basis, I was taught that emptiness—the immeasurable, indefinable potential that is the background of all phenomena, uncreated and unaffected by changes in causes or conditions—is *absolute reality*. And since natural mind is emptiness, completely open and unlimited by any sort of nameable or definable characteristics, nothing anyone thinks or says about phenomena and nothing I think or say about phenomena can truly be said to define its true nature.

In other words, absolute reality cannot be expressed in words, images, or even symbolism of mathematical formulas. I've heard that a number of religions also understand that the nature of the absolute cannot be expressed in these ways, and refuse to describe the absolute in names or images. On this point, at least, Buddhism agrees: The absolute can only be comprehended through experience.

At the same time, it would be absurd to deny that we live in a world where things appear, change, and disappear in space and time. People come and go; tables break and chip; someone drinks a glass of water, and the water is gone. In Buddhist terms, this level of endlessly changing experience is known as *relative reality*—relative, that is, compared with the unchanging and indefinable condition of absolute reality.

So while it would be foolish to pretend that we don't experience things like tables, water, thoughts, and planets, at the same time we can't say that any of these things inherently exists in a complete, self-sufficient, independent way. By definition, anything that inherently exists must be permanent and unchanging. It can't be broken down into smaller parts or affected by shifts in causes and conditions.

That's a nice, intellectual description of the relationship between absolute and relative reality. But it doesn't really provide the intuitive or, as we would say today, gut-level understanding needed to really grasp that relationship. When pressed by his students to explain the relationship between absolute and relative reality, the Buddha often resorted to the example of dreams, pointing out that our experiences in waking life are similar to the experiences we have in dreams. The dream examples he used naturally involved things that were relevant to the students of his day: cows, grain, thatched roofs, and mud walls.

I'm not sure those examples would have the same impact on people living in the twenty-first century. So, when I teach, I tend to use examples relevant to the people I'm talking to. For example, suppose you're the type of person who really loves cars. You'd probably feel thrilled to dream that someone has given you a brand-new car without your hav-

ing to spend a penny to get it. The "dream you" would be happy to receive the "dream car," happy to drive it, and happy to show it off to
everyone that you know.

But suppose in the dream you're driving along when suddenly another
car smashes into you. The front of your car is completely ruined and
you've broken one of your legs. In the dream, your mood would probably
shift immediately from happiness to despair. Your car's been ruined, you
don't have any "dream insurance," and your broken leg is causing
tremendous pain. You might even begin to cry in the dream, and when
you wake up your pillow might be wet with tears.

Now I'm going to ask a question, but not a difficult one.

Is the car in the dream real or not?

The answer, of course, is that it is not. No engineers designed the car,
and no factory built it. It isn't made of the various parts that constitute
an actual car, or of the molecules and atoms that make up each of the
different parts of a car. Yet, while dreaming, you experience the car as
something quite real. In fact, you relate to everything in your dreams as
real, and you respond to your experiences with very real thoughts and
emotions. But, no matter how real your dream experiences may seem,
they can't be said to exist inherently, can they? When you wake up, the
dream ceases and everything you perceived in the dream dissolves into
emptiness: the infinite possibility for anything to occur.

The Buddha taught that, in the same way, every form of experience
is an appearance arising from the infinite possibility of emptiness. As
stated in the *Heart Sutra,* one of the most famous of the Buddha's
teachings:

> Form is emptiness.
> Emptiness is form.
> Emptiness is nothing other than form.
> Form is nothing other than emptiness.

In modern terms, you might say:

> A dream car is a not-inherently-real car.
> A not-inherently-real car is a dream car.

A dream car is nothing other than a not-inherently-real car.
A not-inherently-real car is nothing other than a dream car.

Of course, it may be argued that the things you experience in wak-ing life and the events you experience in a dream can't logically be compared. After all, when you wake from a dream, you don't *really* have a broken leg or a wrecked car in the driveway. If you got into an accident in waking life, though, you might find yourself in the hospital and facing thousands of dollars' worth of damage to your car.

Nevertheless, the basis of your experience is the same in dreams and in waking life: thoughts, feelings, and sensations that vary accord-ing to changing conditions. If you bear this comparison in mind, what-ever you experience in waking life begins to lose its power to affect you. Thoughts are just thoughts. Feelings are just feelings. Sensations are just sensations. They come and go in waking life as quickly and easily as they do in dreams.

Everything you experience is subject to change according to changing conditions. If even a single condition is changed, the form of your expe-rience will change. Without a dreamer, there would be no dream. With-out the mind of the dreamer, there would be no dream. If the dreamer were not sleeping, there would be no dream. All these circumstances have to come together in order for a dream to occur.

AN EXERCISE IN EMPTINESS

The mind is empty in essence.
Although empty, everything constantly arises in it.
  —The Third Gyalwang Karmapa,
    Song of Karmapa: The Aspiration of the Mahamudra of True Meaning,
    translated by Erik Pema Kunsang

Intellectual understanding of emptiness is one thing; direct experience is another. So let's try another exercise, a little different from the ones described in previous chapters. This time you'll look at your thoughts, emotions, and sensations very closely, as they arise out of emptiness, momentarily appear as emptiness, and dissolve back into emptiness. If no thoughts, feelings, or sensations come up for you, just make them up, as many as you can, very quickly, one after another. The main point of the exercise is to observe as many forms of experience as you can. If you don't observe them, they'll just slip away unnoticed. Don't lose any of the thoughts, feelings, or sensations without having observed them.

Begin by sitting up straight, in a relaxed position, and breathing normally. Once you're settled, start to observe your thoughts, emotions, and sensations very clearly. Remember, if nothing comes up for you, just start gibbering away in your mind. Whatever you perceive—pain, pressure, sounds, and so on—observe it very clearly. Even ideas like "This is a good thought," "This is a bad thought," "I like this exercise," or "I hate this exercise" are thoughts you can observe. You can even observe something as simple as an itch. To get the full effect, you'll want to continue this process for at least a minute.

Are you ready? Okay, then, go!

Watch the movements of your mind. . . .

Watch the movements of your mind. . . .

Watch the movements of your mind. . . .

Now stop.

The point of the exercise is simply to watch everything that passes through your awareness as it arises out of emptiness, momentarily appears, and dissolves back into emptiness again—a movement like the rising and falling of a wave in a giant ocean. You don't want to block your thoughts, emotions, and so on; nor do you want to chase after them. If you chase after them, if you let them lead you, they begin to define you, and you lose your ability to respond openly and spontaneously in the present moment. On the other hand, if you attempt to block your thoughts, your mind can become quite tight and small.

This is an important point because many people mistakenly believe meditation involves deliberately stopping the natural movement of thoughts and emotions. It's possible to block this movement for a little while and even achieve a fleeting sense of peace—but it's the peace of a zombie. A completely thoughtless, emotionless state is a state devoid of discernment or clarity.

If you practice allowing your mind just to be as it is, however, your mind will eventually settle down on its own. You'll develop a sense of spaciousness, while your ability to experience things clearly, without bias, will gradually increase. Once you begin to watch these thoughts, emotions, and so on come and go with awareness, you'll start to recognize that they are all relative phenomena. They can only be defined by their relation to other experiences. A happy thought is distinguished by its difference from an unhappy thought, just as a tall person may be distinguished as "tall" only in relation to someone who is shorter. By himself, that person is neither tall nor short. Similarly, a thought or a feeling can't, in itself, be described as positive or negative except through comparison with other thoughts. Without this kind of comparison, a thought, a feeling, or a perception is just what it is. It has no inherent qualities or characteristics, and can't be defined in itself except through comparison.

THE PHYSICS OF EXPERIENCE

> *Physical objects do not exist in space, but are spatially extended.*
> *In this way the concept of "empty space" loses its meaning.*
>
> — ALBERT EINSTEIN,
> *Relativity,* 15th edition

In my conversations with modern scientists, I've been struck by a number of similarities between the principles of quantum mechanics and the Buddhist understanding of the relationship between emptiness and appearance. Because the words we used were different, it took me quite a while to recognize that we were talking about the same thing—phenomena unfolding moment by moment, caused and conditioned by an almost infinite number and variety of events.

In order to appreciate these similarities, I found it important to understand something about the principles of classical physics, the foundation on which quantum mechanics is built. "Classical physics" is a general term that describes a set of theories about the workings of the natural world based on the insights of the seventeenth-century genius Sir Isaac Newton and the scientists who contributed to his understanding and followed in his footsteps. In terms of classical physics, the universe was understood as a giant, orderly machine. According to this "machine model," if one knew the location and *velocity*—that is, the speed and direction of its movement—of every particle in the universe and the forces between them at a particular instant of time, then it would be possible to predict the position and velocity of every particle in the universe at any future time. Similarly, one could figure out the complete past history of the universe from a complete description of its present state. The history of the universe could be understood as a giant web of histories of individual particles, connected by absolute, knowable laws of cause and effect.

The laws and theories of classical physics, however, were based in large part on observations of large-scale phenomena, such as the movements of stars and planets and interactions among material objects on earth. But technological advances in the nineteenth and twentieth centuries enabled scientists to study the behavior of phenomena on smaller and smaller scales, and their experiments—which formed the basis of quantum mechanics (the fundamental framework of modern physics)—began to show that at extremely small scales, material phenomena didn't behave in the nice, orderly, predictable fashion described by classical physics.

One of the most perplexing aspects of these experiments involved the revelation that what we ordinarily consider "matter" may not be as solid and definable as once was believed. When observed on a subatomic level, "matter" behaves rather strangely, sometimes exhibiting properties commonly associated with material particles and sometimes appearing as nonmaterial "waves" of energy. As I understand it, these particle/waves can't be defined simultaneously in terms of location and velocity. So the classical notion of describing the state of the universe in terms of the locations and velocities of particles falls apart.

Just as quantum mechanics developed over time from the laws of classical physics, in a similar sort of way the Buddha's description of the nature of experience evolved gradually, with each insight building upon the previous one according to the level of understanding of those who heard them. These teachings are historically divided into three sets, referred to as the "Three Turnings of the Wheel of Dharma." The Sanskrit word *dharma* in this sense means "the truth," or more simply, "the way things are." The Buddha gave his first set of teachings in an open space known as the Deer Park near Varanasi, a district in India now known as Benares. This first set of teachings described the relative nature of reality based on observable physical experience. The teachings of the first turning are often summed up in a series of statements commonly known as "The Four Noble Truths," but which may be more accurately described as "Four Pure Insights into the Way Things Are." These four insights may be summarized as follows:

1. Ordinary life is conditioned by suffering.
2. Suffering results from causes.
3. The causes of suffering can be extinguished.
4. There is a simple path through which the causes of suffering can be extinguished.

In the second and third turnings, the Buddha began to describe the characteristics of absolute reality. The second turning—which was given on Vulture's Peak, a mountain located in the northeastern Indian state of Bihar—focused on the nature of emptiness, loving-kindness, compassion, and *bodhicitta*. (Bodhicitta is a Sanskrit word that is often translated as the "mind" or "heart" of awakening.) The third turning of the wheel, in which the Buddha described the fundamental characteristics of Buddha nature, was given in various places around India.

On their own, the three turnings of the wheel are fascinating in terms of what they tell us about the nature of the mind, the universe, and the ways in which the mind perceives experience. But they also serve to clarify ideas that arose among the Buddha's earliest followers. After the Buddha passed away, his followers didn't always agree on the exact interpretation of what he'd said; some of them may not have heard all three turnings of the wheel. The disagreements between them were only natural, since, as the Buddha repeatedly stressed, the

essence of what he taught couldn't be grasped by intellectual understanding alone, but could only be realized through direct experience.

Those who had learned only the teachings of the first turning of
the wheel developed two schools of thought, the Vaibhasika and
the Sautantrika views, according to which infinitesimally small
particles—known in Tibetan as *dul-tren* or *dul-tren-cha-may*, which
may be roughly translated, respectively, as "smallest particles" and "indivisible particles"—were understood to be absolutely "real," in the
sense that they were complete in themselves, unable to be broken
down into smaller parts. These fundamental particles were considered
the essential building blocks of all phenomena. They could never be
dissolved or lost, only converted to different forms. For example, the
dul-tren-cha-may of wood wasn't lost when a log was set on fire, but
was merely converted into smoke or flame—a point of view not unlike
the law of "conservation of energy," a basic principle of physics that
holds that energy can neither be created nor destroyed, but only converted into other forms. For example, the chemical energy in gasoline
can be converted into the mechanical energy that moves a car.

At this point you may be wondering what the development of
modern physics has to do with attaining personal happiness. But if
you'll bear with me for a while, the relationship will become clear.

The Buddha's later teachings demonstrated that the simple fact that
infinitesimally small particles *could* be converted—as Albert Einstein
would prove centuries later through his famous equation $E = mc^2$,
which in very basic terms describes particles as little packets of
energy—indicated that a dul-tren or dul-tren-cha-may was, in fact, a
transitory phenomenon and consequently could not be considered fundamentally, or absolutely, "real."

To use an everyday example, think of water. Under very cold conditions, water turns to ice. At room temperature, water is liquid. When
heated, it becomes steam. In laboratory experiments, the molecules of
water can be separated into hydrogen and oxygen atoms, and when
these atoms are examined more closely, they consist of smaller and
smaller subatomic particles.

There is an interesting parallel between the Vaibhasika and the

Sautantrika views and the classical school of physics. According to classical physics—and I'm probably oversimplifying the case in order to make the ideas easier to grasp—the basic elements of matter, as well as large entities like stars, planets, and human bodies, can be described in terms of precisely measurable properties such as location and velocity, and move in nicely predictable ways through space and time in perfect coordination with certain forces such as gravity and electricity. The classical interpretation still works very well in terms of predicting the behavior of large-scale phenomena, like the movements of planets.

As explained to me, however, advances in technology in the nineteenth century began to provide physicists with the means to observe material phenomena in microscopic detail. In the early twentieth century the British physicist J. J. Thomson pursued a set of experiments that led him to discover that the atom was not a solid entity, but was instead composed of smaller particles—most notably electronically charged particles called *electrons*. Building on Thomson's experiments, the physicist Edward Rutherford devised a model of the atom—familiar to most Westerners who have taken a high school chemistry or physics class—as a kind of miniature solar system composed of electrons that revolved around a central core of the atom called the *nucleus*.

The problem with Rutherford's "solar system" model of the atom was that it didn't account for the observed fact that atoms always radiate light of certain characteristic energies when heated up. The set of energy levels, which is different for each type of atom, is commonly referred to as the atom's *spectrum*. In 1914, Niels Bohr realized that if the electrons inside an atom were treated as waves, the atom's energy spectrum could be precisely explained. This was one of the great early successes of quantum mechanics, and forced the scientific world to begin taking this strange new theory seriously.

At about the same time, however, Albert Einstein demonstrated that it was possible to describe light not as waves, but as particles, which he called *photons*. When photons were directed at a metal plate, they accelerated the activity of electrons, producing electricity. Following up on Einstein's discovery, a number of physicists began experiments that showed that all forms of energy might conceivably be described in terms of particles—a perspective very similar to the Vaibhasika point of view.

As modern physicists continue to study the world of subatomic phenomena, they're still confronted by the problem that subatomic phenomena—what we might call the building blocks of "reality" or "experience"—sometimes behave like particles and sometimes like waves. Thus, they can only determine the *probability* that a subatomic entity will exhibit certain properties or behave in a certain way. While there appears to be no doubt that quantum theory is accurate in terms of practical application—as demonstrated in the development of lasers, transistors, supermarket scanners, and computer chips—the quantum explanation of the universe remains a rather abstract mathematical description of phenomena. But it's important to remember that mathematics is a symbolic language—a type of poetry that uses numbers and symbols instead of words to convey a sense of reality underlying our conventional experience.

THE FREEDOM OF PROBABILITY

Fresh awareness of whatever arises . . . is sufficient.

— The Ninth Gyalwang Karmapa,
Mahāmudrā: The Ocean of Definitive Meaning,
translated by Elizabeth M. Callahan

In his early teachings, the Buddha addressed the problem of suffering in terms of fixation on an inherently existing or absolutely "real" level of experience (including a belief in an inherently real self and the inherently real existence of material phenomena). Later, as his audience became more sophisticated, he began to address emptiness and Buddha nature more directly. Similarly, the ideas of classical physicists regarding the nature and behavior of material objects were gradually redefined and updated by the efforts of scientists of the late nineteenth century.

As mentioned earlier, modern scientists' observations of matter on the subatomic level led them to recognize that elements of the subatomic world sometimes behaved very nicely as "thinglike" particles when observed under certain experimental conditions, but when observed under other conditions they behaved more like waves. These observations of "wave-particle duality" marked, in many ways, the birth of the "new" physics of quantum mechanics.

I imagine that this peculiar behavior was probably not altogether comfortable to the scientists who first observed it. To use a somewhat simple analogy, imagine someone you thought you knew very well treated you like a best friend one moment, and half an hour later looked at you as though he or she had never seen you before. You might call this kind of behavior "two-faced."

On the other hand, it must have been very exciting, since direct observation of the behavior of matter opened up a whole new world of investigation—quite similar to the world that opens up to us when we begin to actively engage in observing the activity of our own minds. There's so much to see, and so much to learn!

With their customary diligence, physicists of the early twentieth century went "back to the drawing board" in order to explain the behavior of the wavelike nature of particles. Building on Niels Bohr's picture of the wavelike nature of electrons inside atoms, they eventually arrived at a new description of the subatomic world, which, in very detailed, mathematical terms, describes how every particle in the known universe can be understood as a wave, and every wave as a particle. In other words, the particles that make up the larger material universe can be seen from one perspective as "things" and from another as occurrences extending through time and space.

So what does physics have to do with being happy? We like to think of ourselves as solid, distinct individuals with well-defined goals and personalities. But if we look honestly at the discoveries of modern science, we have to admit that this view of ourselves is at best incomplete.

The teachings of the Buddha are often grouped into two categories: the teachings on wisdom, or theory, and the teachings on method, or practice. The Buddha himself often compared these categories to the wings of a bird. In order to fly, a bird needs two wings. The "wing" of wisdom is necessary because without at least some idea of what you're aiming for, the "wing" of practice flaps pretty much uselessly. People who go to the gym, for example, have at least a rough idea of what they want to gain by sweating on the treadmill or lifting weights. The same principle applies to the effort to directly recognize our inborn capacity for happiness. We need to know where we're going in order to get there.

Modern science—specifically quantum physics and neuroscience—offers an approach to wisdom in terms that are at once more acceptable and more specifically demonstrable to people living in the twenty-first century than are the Buddhist insights into the nature of reality gained through subjective analysis. It not only helps to explain why the Buddhist practices work in terms of hard, scientific analysis, but also provides fascinating insights into the Buddhist understanding of dul-tren-cha-may, the momentary phenomena that appear and disappear in an instant according to changes in causes and conditions. But we have to look deeper into the realm of science to discover some of these parallels.

5

THE RELATIVITY OF PERCEPTION

*The primordial purity of the ground completely transcends
words, concepts, and formulations.*

> —JAMGÖN KONGTRUL, *Myriad Worlds*,
> translated and edited by the International Committee
> of Kunkhyab Chöling

THE DEFINITION OF emptiness as "infinite possibility" is a basic description of a very complicated term. A subtler meaning, which might have been lost on early translators, implies that whatever arises out of this infinite potential—whether it's a thought, a word, a planet, or a table—doesn't truly exist as a "thing" in itself, but is rather the result of numerous causes and conditions. If any of those causes or conditions are changed or removed, a different phenomenon will arise. Like the principles outlined in the second turning of the wheel of Dharma, quantum mechanics tends to describe experience in terms not simply of a single possible chain of events leading to a single result, but rather of probabilities of events and occurrences—which, in an odd way, is closer to the Buddhist understanding of absolute reality, in which a variety of outcomes are theoretically possible.

INTERDEPENDENCE

Whatever depends on conditions is explained to be empty. . . .

—*Sutra Requested by Madröpa,*
translated by Ari Goldfield

To use a simple example, imagine two different chairs: one that has four sturdy legs and one that has two good legs and two cracked ones. If you sit in the chair that has four good legs, you'll be very comfortable. Sit in the other one and you'll end up on the floor. On a superficial level, they can both be said to be "chairs." But your experience of each "chair" will be unmistakably different because the underlying conditions are not the same.

This coming together of different causes is known, in Buddhist terms, as *interdependence*. We can see the principle of interdependence at work all the time in the world around us. A seed, for example, carries within itself the potential for growth, but it can only realize its potential—that is, become a tree, a bush, or a vine—under certain conditions. It has to be planted, watered, and given the proper amount of light. Even under the right conditions, whatever grows depends on the kind of seed planted. An apple seed won't grow into an orange tree, nor will an orange seed become a tree that suddenly sprouts apples. So, even within a seed, the principle of interdependence applies.

Similarly, the choices we make in our daily lives do have a relative effect, setting in motion causes and conditions that create inevitable consequences in the domain of relative reality. Relative choices are like stones tossed in a pond. Even if the stone doesn't go very far, wherever it falls, concentric ripples will spread out from the area where the stone hits. There's no way for this *not* to happen (unless, of course, your aim is really bad and you miss the pond altogether and send a stone sailing through your neighbor's window, in which case a whole different set of consequences will occur).

In the same way, your ideas about yourself—"I'm not good enough," "I'm too fat," or "I made a horrible mistake yesterday"—are based on prior causes and conditions. Maybe you didn't sleep well the night before. Maybe someone said something you didn't like earlier in the day. Or maybe you're just hungry and your body is crying out for vitamins or minerals that it needs to function properly. Something as simple as a lack of water can cause fatigue, headaches, and an inability to concentrate. Any number of things can determine the nature of relative experience without changing the absolute reality of who you are.

When I was being examined by neuroscientists at the laboratory in Wisconsin, I asked a lot of questions about how modern scientists understand perception. Buddhists have their own theories, but I was curious about the Western scientific point of view. What I learned was that from a strictly neuroscientific standpoint any act of perception requires three essential elements: a stimulus—such as a visual form, a sound, a smell, a taste, or something we touch or that touches us; a sensory organ; and a set of neuronal circuits in the brain that organize and make sense of the signals received from the sense organ.

Using visual perception of a banana as an example, the scientists I spoke with explained that the optic nerves—the sensory neurons in the eye—first detect a long yellow curved thing, which maybe has a brown spot at either end. Excited by this stimulus, the neurons start firing off messages to the *thalamus,* a neuronal structure located at the very center of the brain. The thalamus is something like a central switchboard, like the kind portrayed in old movies, where sensory messages are sorted before being passed to other areas of the brain.

Once the messages from the optic nerves are sorted by the thalamus, they're sent to the limbic system, the region of the brain chiefly responsible for processing emotional responses and sensations of pain and pleasure. At this point our brains make a sort of immediate judgment on whether the visual stimulus—in this case the long yellow curved thing with brown spots at either end—is a good thing, a bad thing, or something neutral. Like the feeling we sometimes get in the presence of other people, we tend to refer to this immediate response as a "gut reaction," though it doesn't occur entirely in the stomach. It's just a lot easier to use this shorthand description than to go into details like "a stimulation of neurons in the limbic region."

As this information is processed in the limbic area, it is simultaneously passed "higher up" to the regions of the neocortex, the mainly analytical region of the brain, where it's organized into patterns—or, more specifically, concepts—that provide the guide or map we use to navigate the everyday world. The neocortex evaluates the pattern and arrives at the conclusion that the object that stimulated our optic nerve cells is, in fact, a banana. And if the neocortex has already created the pattern or concept "banana," it offers up all sorts of associated details based on past experiences—for example, what a banana tastes like, whether we like the taste or not, and all sorts of other details related to our concept of a banana, all of which enable us to decide how to respond with greater precision to the object we see as a banana.

What I've described is just a bare outline of the process of perception. But even a glimpse of the process provides a clue to how even an ordinary object can become a cause of happiness or unhappiness. Once we've arrived at the stage where we recognize a banana, we're really not seeing the original object anymore. Instead, we're seeing an *image* of it constructed by the neocortex. And this image is conditioned by a huge variety of factors, including our environment, expectations, and prior experiences, as well as the very structure of our neuronal circuitry. In the brain itself, the sensory processes and all these factors can be said to be interdependent in the sense that they continuously influence one another. Because the neocortex ultimately provides the pattern by which we're able to recognize, name, and predict the behavior, or "rules," associated with an object we perceive, it does, in a very profound sense, shape the world for us. In other words, we're not seeing the *absolute* reality of the banana, but rather its relative appearance, a mentally constructed image.

To illustrate this point, during the first Mind and Life Institute conference in 1987, Dr. Livingston described a simple experiment that involved presenting a group of research subjects with the letter T, carefully drawn so that both the horizontal and the vertical segments were exactly equal in length.[1] When asked whether one of the two segments was longer than the other or equal in length, three different responses were given, each based on the subjects' backgrounds. For example, most of the people who lived or had been raised in mainly flat environments, like the Netherlands, tended to see the horizontal (or flat) segment as longer. By contrast, people living or raised in mountainous environments, and therefore more likely to perceive things in terms of up and down, were overwhelmingly convinced that the vertical segment was longer. Only a small group of subjects was able to recognize the two segments as equal in length.

In strictly biological terms, then, the brain is an active participant in the shaping and conditioning of perception. Although scientists would not deny that there is a "real world" of objects beyond the confines of the body, it's generally agreed that even though sensory experiences appear to be very direct and immediate, the processes involved are far more subtle and complex than they appear. As Francisco Varela commented later on in the conference, "It's as if the brain actually makes the world come through in perception."[2]

The brain's active role in the process of perception plays a critical part in determining our ordinary state of mind. And this active role opens the possibility for those willing to undertake certain practices of mental training to gradually change long-standing perceptions shaped by years of prior conditioning. Through retraining, the brain can develop new neuronal connections, through which it becomes possible not only to transform existing perceptions but also to move beyond ordinary mental conditions of anxiety, helplessness, and pain and toward a more lasting experience of happiness and peace.

This is good news for anyone who feels trapped in ideas about the way life is. Nothing in your experience—your thoughts, feelings, or sensations—is as fixed and unchangeable as it appears. Your perceptions are only crude approximations of the true nature of things. Actually, the universe in which you live and the universe in your mind form an integrated whole. As explained to me by neuroscientists, physicists, and psychologists, in a bold effort to describe reality in objective, rational terms, modern science has begun to restore in us a sense of the magic and majesty of existence.

SUBJECTS AND OBJECTS: A NEUROSCIENTIFIC VIEW

> *Dualistic thought is the dynamic energy of mind.*
>
>   —JAMGÖN KONGTRUL, *Creation and Completion,*
>   translated by Sarah Harding

Armed with a bit more information about physics and biology, we can ask some deeper questions about the absolute reality of emptiness and the relative reality of daily experience. For example, if what we perceive is just an image of an object, and the object itself, from the point of view of a physicist, is a whirling mass of tiny particles, then why do we experience something like a table in front of us as solid? How can we see and feel a glass of water on the table? If we drink the water, it seems real and tangible enough. How can that be? If we don't drink water, we'll be thirsty. Why?

To begin with, the mind engages in many ways in a process that is known as *dzinpa,* a Tibetan word that means "grasping." Dzinpa is the tendency of mind to fixate on objects as inherently real. Buddhist training offers an alternative approach to experiencing life from an essentially fear-based perspective of survival in favor of experiencing it as a parade of odd and wonderful events. The difference can be demonstrated through a simple example. Imagine that I'm holding my *mala* (a string of prayer beads similar to a rosary) in my hand with my palm turned downward. For this example, the mala represents all the possessions people usually feel they need: a nice car, fine clothes, good food, a well-paying job, a comfortable home, and so on. If I hold my mala tightly, some part of it always seems to escape my grasp and hang outside my hand. If I try to grasp the loose part, a longer bit of the mala falls through my fingers; and if I try to grasp that, an even longer piece slips through. If I continue this process, I'll eventually lose my grasp on the entire mala. If, however, I turn my palm upward, and allow the mala to simply rest in my open palm, nothing falls through. The beads sit in my hand loosely.

To use another example, imagine you're sitting in a room full of people looking at a table at the front of the room. Your tendency is to relate to the table as a thing in itself, a completely whole, self-contained object, independent of subjective observation. But a table has a top, legs, sides, a back, and a front. If you remember that it's made up of these different parts, can you really define it as a singular object?

In their exploration of the "conductor-less" brain, neuroscientists have discovered that the brains of sentient beings have evolved specifically to recognize and respond to patterns. Among the billions of neurons that make up the human brain, some neurons are specifically adapted to detect shapes, while others are dedicated to detecting colors, smells, sounds, movements, and so on. At the same time, our brains are endowed with mechanisms that enable us to extract what neuroscientists call "global," or patternlike, relationships.

Consider the familiar example of a little group of visual symbols, called emoticons, often used in e-mail messages: :-). This group is easily recognized as a "smiley face," with two eyes ":," a nose "-," and a mouth ")." If, however, these three objects were rearranged as) - :, the brain wouldn't recognize a pattern and would merely interpret the shapes as random dots, lines, and curves.

Neuroscientists I've spoken with have explained that these pattern-recognition mechanisms operate almost simultaneously with the neuronal recognition of shapes, colors, and so on through *neuronal synchrony*—which, in very simple terms, may be described as a process in which neurons across widely separated areas of the brain spontaneously and instantaneously communicate with one another. For instance, when the shapes :-) are perceived in this precise formation, the corresponding neurons signal one another in a spontaneous yet precisely coordinated fashion that represents recognition of a specific pattern. When no pattern is perceived, the corresponding neurons signal one another randomly.

This tendency to identify patterns or objects is the clearest biological illustration of dzinpa I have so far encountered. I suspect it evolved as some sort of survival function, since the ability to discriminate among harmful, beneficial, and neutral objects or events would be quite handy! As I'll explain later on, clinical studies indicate that the practice of meditation extends the mechanism of neuronal synchrony to a point where the perceiver can begin to recognize consciously that his or her mind and the experiences or objects that his or her mind perceives are one and the same. In other words, the practice of meditation over a long period dissolves artificial distinctions between subject and object—which in turn offers the perceiver the freedom to determine the quality of his or her own experience, the freedom to distinguish between what is real and what is merely an appearance.

Dissolving the distinction between subject and object, however, doesn't mean that perception becomes a great big blur. You still continue to perceive experience in terms of subject and object, while at

the same time recognizing that the distinction is essentially conceptual. In other words, the perception of an object is not different from
the mind that perceives it.

Because this shift is difficult to grasp intellectually, in order to
develop some understanding, it's necessary to resort once again to the
analogy of a dream. In a dream, if you recognize that what you're experiencing is just a dream, then you also recognize that whatever you experience in the dream is merely occurring in your own mind.
Recognizing this, in turn, frees you from the limitations of "dream
problems," "dream suffering," or "dream limitations." The dream still
continues, but recognition liberates you from whatever pain or un-
pleasantness your dream scenarios present. Fear, pain, and suffering
are replaced by a sense of almost childlike wonder: "Wow, look what
my mind is capable of producing!"

In the same way, in waking life, transcending the distinction between subject and object is equivalent to recognizing that whatever
you experience is not separate from the mind that experiences it. Waking life doesn't stop, but your experience or perception of it shifts from
one of limitation to one of wonder and amazement.

THE GIFT OF UNCERTAINTY

> *When the mind is without reference point, that is mahāmudrā.*
> —Tilopa, *Ganges Mahāmudrā*,
> translated by Elizabeth M. Callahan

If we return to the example of looking at a table, we can say that even
on a normally observable level, a table is in a constant state of change.
Between yesterday and today, some of the wood might have broken off
or some of the paint may have chipped. If we look at the table from the
perspective of a physicist, on a microscopic level we'd see that the
wood, paint, nails, and glue that make up the table are composed of
molecules and atoms made up of rapidly moving particles that fluctu-
ate through the vastness of subatomic space.

On this subatomic level, physicists encounter an interesting problem: When they seek to measure the precise location of a particle in subatomic space, they can't measure its velocity with 100 percent accuracy; and when they try to measure a particle's velocity, they can't precisely identify its location. The problem of simultaneously measuring the exact position and velocity of a particle is known as Heisenberg's Uncertainty Principle, named after Werner Heisenberg, one of the founders of quantum mechanics.

Part of the problem, as explained to me, is that in order to "see" the position of a subatomic particle, physicists must shine a brief pulse of light at it, which supplies the particle with an extra "kick" of energy and changes the particle's rate of movement. On the other hand, when physicists try to measure the velocity of a particle, they do so by measuring the changes in frequency of light waves beamed at the particle as it moves—similar to the way traffic police use the frequency of radar waves to measure the speed of a car. Thus, depending on the experiment scientists are performing, they gain information about one or the other property of the particle. Put very simply, the results of an experiment are conditioned by the nature of the experiment—that is, by the questions asked by the scientists who set up and observe the experiment.

If you consider this paradox as a way of describing human experience, you can see that just as the qualities ascribed to a particle are determined by the particular experiment scientists perform on it, in a related fashion, everything we think, feel, and perceive is conditioned by the mental habits we bring to it.

Modern physics has indicated that our understanding of material phenomena is limited to some extent by the questions we ask of it. At the same time, the uncertainty of predicting exactly how and where a particle may appear in the subatomic universe represents a certain freedom in determining the nature of our experience.

CONTEXT: A COGNITIVE PERSPECTIVE

> *Our life is shaped by our mind.* . . .
> —The Dhammapada,
>   translated by Eknath Easwaran

Buddhist practice guides us very gradually to let go of habitual assumptions and experiment with different questions and different points of view. Such a shift in perspective isn't as difficult as it might seem. During a conversation I had in Nepal with a student of mine who works in the field of cognitive psychology, I learned that the ability to shift our way of seeing things is a basic function of the human mind. In cognitive psychological terms, the meaning of whatever information we receive is determined in large part by the *context* in which we view it. The different levels of context seem to bear a striking resemblance to the different ways of observing reality in terms of quantum mechanics.

For example, if we look at the words

M I N G Y U R   R I N P O C H E

it's possible to interpret their meaning in a variety of different ways, including the following:

- an arrangement of lines and spaces
- a group of letters
- just a name
- a reference to a specific person we know
- a reference to a specific person we don't know

There are probably more levels of interpretation, but we can stick with these five for our example.

What's interesting is that none of the possible interpretations invalidates any of the others. They simply represent different levels of meaning based on context, which, in turn, is based largely on experience.

If you happen to know me personally, for example, you can look at the words "Mingyur Rinpoche" and think, *Oh, yeah, he's that short Tibetan guy with glasses who goes around in red robes telling everyone that tables don't absolutely exist.*

If you didn't know me or anything about me, but only saw the words in a magazine or newspaper article about Tibetan Buddhist teachers, "Mingyur Rinpoche" would just be the name of one of those short Tibetan guys with glasses who go around in red robes telling everyone that tables don't absolutely exist. If you were unfamiliar with the Western alphabet, you might recognize "Mingyur Rinpoche" as a group of letters, but you wouldn't know what they meant, or whether they referred to a name or a place. And if you had no familiarity with alphabets at all, the words would just be an odd, possibly interesting collection of lines and circles that might or might not have any meaning.

So, when I'm talking about abandoning everyday logic and applying a different perspective to our experience, what I'm suggesting is that as you start to look more closely at things, you can begin to appreciate how very difficult it is to pinpoint their *absolute* reality. You can begin to see that you've invested things with permanence or self-existence as a result of the context in which you've viewed them; and if you practice seeing yourself and the world around you from a different point of view, then your perception of yourself and the world around you will shift accordingly.

Of course, changing your perceptions and expectations about the material world requires not only effort, but also time. So, in order to get past this obstacle and truly begin to experience the freedom of emptiness, you have to learn to look at time itself in a different light.

THE TYRANNY OF TIME

> *The past is imperceptible, the future is imperceptible, and the present is imperceptible. . . .*
>
> —Sutras of the Mother,
> translated by Ari Goldfield

If you look at your experience from the point of view of time, you can say that tables, glasses of water, and so on do indeed exist in time—but only from a relative perspective. Most people tend to think of time in terms of past, present, and future. "I *went* to a boring meeting." "I'm *in* a boring meeting." "I *have to go* to a boring meeting." "I *fed* my children this morning." "I'm *feeding* my children lunch right now." "Oh no, *I have to make dinner* for my children and there's nothing in the refrigerator, so I have to go to the store as soon as I get out of this boring meeting!"

Actually, though, when you think of the past, you're merely recalling an experience that has already happened. You're out of the meeting. You've fed your children. You've finished your shopping. The past is like a seed that's been burned in a fire. Once it's burned to ashes, there's no more seed. It's only a memory, a thought passing through the mind. The past, in other words, is nothing more than an idea.

Likewise, what people tend to call "the future" is an aspect of time that hasn't yet occurred. You wouldn't talk about a tree that hasn't been planted as though it were a solid, living object, because you have no context for talking about it; nor would you talk about children who haven't yet been conceived the way you would about people you're dealing with here and now. So the future, too, is just an idea, a thought passing through your mind.

So what are you left with as an actual experience?

The present.

But how is it even possible to define "the present"? A year is made up of twelve months. Every day of each month is made up of twenty-four hours. Every hour is made up of sixty minutes; every minute is made up of sixty seconds; and every second is made up of microseconds and nanoseconds. You can break down the present into smaller and smaller increments, but between the instant of present experience and the instant you identify that instant as "now," the moment has already passed. It's no longer *now*. It's *then*.

The Buddha intuitively understood the limitations of the ordinary human conception of time. In one of his teachings he explained that from a relative point of view the division of time into distinct periods of duration such as an hour, a day, a week, and so on, might have a certain degree of relevance. But from an absolute perspective, there's really no difference between a single instant of time and an eon. Within an eon there can be an instant; within an instant there can be an eon. The relationship between the two periods would not make the instant any longer or the eon any shorter.

He illustrated this point through a story about a young man who came to a great master in search of a profound teaching. The master agreed, but suggested the young man first have a cup of tea. "After that," he said, "I'll give you the profound teaching you've come looking for."

So the master poured a cup of tea, and as the student brought it to his mouth, the cup of tea transformed itself into a broad lake surrounded by mountains. As he stood beside the lake, admiring the beauty of the scene, a girl stepped from behind him and approached the lake to gather water in a pail. For the young man, it was love at first sight, and as the girl looked at the young man standing beside the lake, she fell in love with him, too. The young man followed her back to her home, where she lived with her aged parents. Gradually the girl's parents grew to be fond of the young man, and he of them, and it was eventually agreed that the two young people should marry.

After three years, the couple's first child was born, a son. A few years later a daughter was born. The children grew up happy and strong, until one day, at the age of fourteen, the son fell ill. None of the medicines prescribed for him cured his illness. Within a year he was dead.

Not long afterward, the couple's young daughter went to gather wood in the forest, and while she was busy with her task, she was attacked and killed by a tiger. Unable to overcome her sorrow over losing both her children, the young man's wife eventually decided to drown herself in the nearby lake. Distraught over the loss of their daughter and their grandchildren, the girl's parents stopped eating, eventually starving themselves to death. Having lost his wife, his children, and his parents-in-law, the young man began to think that he might as well die himself. He walked to the edge of the lake, determined to drown himself.

Just as he was about to jump into the water, however, he suddenly found himself back in the master's house, holding his cup of tea up to his lips. Though he had lived an entire lifetime, hardly an instant had passed; the cup was still warm in his hands and the tea was still hot.

He looked across the table at the teacher, who nodded, saying,

"Now you see. All phenomena proceed from the mind, which is emptiness. They do not truly exist except in the mind, but they are not nothingness. There is your profound teaching."

From a Buddhist perspective, the essence of time, like the essence of space and the objects that move around in space, is emptiness. At a certain point, any attempt to examine time or space in terms of smaller and smaller intervals finally breaks down. You can experiment with your perception of time through meditation, trying to look at time in smaller and smaller increments. You can try to examine time in this way until finally you reach a point where you can't name or define anything anymore. When you reach that point, you enter an experience that is beyond words, beyond ideas, beyond concepts.

"Beyond ideas and concepts" doesn't mean that your mind becomes as empty as an eggshell or as dull as a stone. Actually, quite the opposite occurs. Your mind becomes more vast and open. You can still perceive subjects and objects, but in a more illusory way: You recognize them as concepts, not as inherently or objectively real entities.

I've spoken to many scientists about whether ideas parallel to the Buddhist view of time and space could be found among modern theories and discoveries. Though many ideas were suggested to me, nothing seemed to fit quite precisely until I was introduced to the theory of *quantum gravity*, an examination of the fundamental nature of space and time that explores such basic questions as "What are space and time made of? Do they exist absolutely or do they emerge from something more fundamental? What do space and time look like on very small scales? Is there a smallest possible length or unit of time?"

As it has been explained to me, in most branches of physics, space and time are treated as though they were infinite, uniform, and perfectly smooth: a static background through which objects move and events happen. This is a very workable assumption for examining the nature and properties of both large bodies of matter and subatomic particles. But when it comes to examining time and space themselves, the situation becomes very different.

At the level of ordinary human perception, the world looks sharp, clear, and solid. A plank supported by four legs appears on the level of ordinary perception quite obviously to be a table. A cylindrically shaped object with a flat bottom and an open top appears quite obviously to be a glass. Or, if it has a handle, maybe we'd call it a cup.

Now imagine looking at a material object through a microscope. You might reasonably expect that by gradually increasing the microscope's level of magnification you'd see a sharper, clearer image of the object's underlying structure. Actually, however, the opposite occurs. As we approach a magnification where we are able to see individual atoms, the world begins to look more and more "fuzzy," and we leave most of the rules of classical physics behind. This is the realm of quantum mechanics, in which, as described earlier, subatomic particles jitter about in all possible ways and pop in and out of existence with increasing frequency.

Continuing to increase the magnification so that we can see smaller and smaller objects, we eventually find that space and time themselves start to jitter—space itself develops tiny curves and kinks that appear and disappear inconceivably fast. This happens at extremely small scales—as small compared to an atom as an atom is compared to the solar system. This state has been called "spacetime foam" by physicists. Think of shaving foam that looks smooth from a distance but close up is composed of millions of tiny bubbles.

Perhaps a better analogy for this state is rapidly boiling water. At even shorter distances and time scales, the water itself boils away, and space and time themselves lose their meaning. At this point, physics itself begins to jitter, because the study of matter, energy, and motion, and the way they relate to one another, cannot even be formulated without an underlying reference to time. At this point, physicists admit, they have no idea how to describe what is left. It is a state that literally includes all possibilities, beyond space and time.

From a Buddhist perspective, the description of reality provided by quantum mechanics offers a degree of freedom to which most people are not accustomed, and that may at first seem strange and even a little frightening. As much as Westerners in particular value the capacity for freedom, the notion that the act of observation of an event can influence the outcome in random, unpredictable ways can seem like too much responsibility. It's much easier to assume the role of the victim and assign the responsibility or blame for our experience to some person or power outside oneself. If we're to take the discoveries of modern science seriously, however, we have to assume responsibility for our moment-by-moment experience.

While doing so may open up possibilities we might never before have imagined, it's still hard to give up the familiar habit of being a victim. On the other hand, if we began to accept responsibility for our experience, our lives would become a kind of playground, offering innumerable possibilities for learning and invention. Our sense of personal limitation and vulnerability would gradually be replaced by a sense of openness and possibility. We would see those around us in an entirely new light—not as threats to our personal security or happiness, but as people simply ignorant of the infinite possibilities of their own nature. Because our own nature is unconstrained by arbitrary distinctions of being "this way" or "that way," or having only certain capabilities and lacking others, then we would be able to meet the demands of any situation in which we might find ourselves.

IMPERMANENCE

Nothing ever lasts. . . .

—Patrul Rinpoche, *The Words of My Perfect Teacher,*
translated by the Padmakara Translation Group

Most people are conditioned by the societies in which they live to apply conceptual labels to the constantly changing stream of mental and material phenomena. For example, when we looked closely at a table, we still instinctively labeled it as a table—even though we saw that it was not a single thing but was actually made up of a number of different parts: a top, legs, sides, a back, and a front. None of these parts could actually be identified as "the table" itself. In fact, "table" was just a name we applied to rapidly arising and dissolving phenomena that merely produced an illusion of something definitely or absolutely real.

In the same way, most of us have been trained to attach the name "I" or "me" to a stream of experiences that confirm our personal sense of self, or what is conventionally referred to as "ego." We feel about ourselves that we're this single entity that continues unchanged over time. In general, we tend to feel we're the same person today as we were yesterday. We remember being teenagers and going to school, and tend to feel that the "I" that we are now is the same "I" that went to school, grew up, moved away from home, got a job, and so on.

But if we look at ourselves in a mirror, we can see that this "I" has changed over time. Maybe we can see wrinkles now that weren't there a year ago. Maybe we're wearing glasses. Maybe we have a different hair color—or no hair at all. On a basic molecular level, the cells in our bodies are always changing, as old cells die off and new ones are generated. We can also examine this sense of "self" the same way we looked at the table, and see that this thing we call "I" is really made up of a number of different parts. It has legs, arms, a head, hands, feet, and internal organs. Can we identify any of these separate parts as definitely "I"?

We might say, "Well, my hand is not me, but it's my hand." But the hand is made up of five fingers, a front, and a back. Each one of them can be broken down even further as nails, skin, bones, and so on. Which of those components can we uniquely identify as our "hand"? We can keep up this line of investigation down to the atomic and subatomic level, and still find ourselves faced with the same problem of being unable to find anything we can definitely identify as "I."

So, whether we're analyzing material objects, time, our "self," or our mind, eventually we reach a point where we realize that our analysis breaks down. At that point our search for something irreducible finally collapses. In that moment, when we give up looking for something absolute, we gain our first taste of emptiness, the infinite, indefinable essence of reality as it is.

As we contemplate the enormous variety of factors that must come together to produce a specific sense of self, our attachment to this "I" we think we are begins to loosen. We become more willing to let go of the desire to control or block our thoughts, emotions, sensations, and so on and begin to experience them without pain or guilt, absorbing their passage simply as manifestations of a universe of infinite possibilities. In so doing, we regain the innocent perspective most of us knew as children. Our hearts open up to others, like flowers blossoming. We become better listeners, more fully aware of everything going on around us, and are able to respond more spontaneously and appropriately to situations that used to trouble or confuse us. Gradually, perhaps on a level so subtle we might not even notice it's happening, we find ourselves awakening to a free, clear, loving state of mind beyond our wildest dreams.

But it takes great patience to learn how to see such possibilities.

In fact, it takes great patience to see.

6

THE GIFT OF CLARITY

All phenomena are expressions of the mind.
> —THE THIRD GYALWANG KARMAPA, *Song of Karmapa:*
> *The Aspiration of the Mahamudra of True Meaning,*
> translated by Erik Pema Kunsang

ALTHOUGH WE COMPARE emptiness to space as a way to understand the infinite nature of the mind, the analogy isn't perfect. Space—at least as far as we know—isn't conscious. From the Buddhist perspective, however, emptiness and awareness are indivisible. You can't separate emptiness from awareness any more than you can separate wetness from water or heat from fire. Your real nature, in other words, is not only unlimited in its potential, but also completely aware.

This spontaneous awareness is known in Buddhist terms as *clarity*, or sometimes as the *clear light of mind*. It's the cognizant aspect of the mind that allows us to recognize and distinguish the infinite variety of thoughts, feelings, sensations, and appearances that perpetually emerge out of emptiness. Clarity operates even when we're not consciously attentive—for instance, when we suddenly think, *I need to eat, I need to leave, I need to stay.* Without this clear light of mind, we wouldn't be able to think, feel, or perceive anything. We wouldn't be able to recognize our own bodies, or the universe or anything that appears in it.

NATURAL AWARENESS

Appearances and mind exist like fire and heat.

—ORGYENPA, quoted in *Mahāmudrā: The Ocean of Definitive Meaning,*
translated by Elizabeth M. Callahan

My teachers described this clear light of mind as self-illuminating—
like the flame of a candle, which is both a source of illumination and
illumination itself. Clarity is part of the mind from the beginning, a
natural awareness. You can't *develop* it the way, for instance, you develop muscles through physical exercise. The only thing you have to do
is acknowledge it, simply *notice* the fact that you're aware. The challenge, of course, is that clarity, or natural awareness, is so much a part
of everyday experience that it's hard to recognize. It's like trying to see
your eyelashes without using a mirror.

So how do you go about recognizing it?

According to the Buddha, you meditate—though not necessarily in
the way most people understand it.

The kind of meditation involved here is, again, a type of "non-
meditation." There's no need to focus on or visualize anything.
Some of my students call it "organic meditation—meditation without
additives."

As in other exercises my father taught me, the way to begin is to sit
up straight, breathe normally, and gradually allow your mind to relax.
"With your mind at rest," he instructed those of us in his little teaching
room in Nepal, "just allow yourself to become aware of all the
thoughts, feelings, and sensations passing through it. And as you
watch them pass, simply ask yourself, 'Is there a difference between
the mind and the thoughts that pass through it? Is there any difference
between the thinker and the thoughts perceived by the thinker?' Continue watching your thoughts with these questions in mind for about
three minutes or so, and then stop."

So there we all sat, some of us fidgeting, some of us tense, but all of
us focused on watching our minds and asking ourselves whether there
was any difference between thoughts and the thinker who thinks the
thoughts.

Since I was just a child, and most of the other students were adults,

I naturally thought they were doing a much better job than I was. But as I watched these thoughts about my own inadequacy pass through my mind, I remembered the instructions, and a funny thing happened. For just a moment, I glimpsed that the thoughts about not being as good as the other students were just thoughts, and the thoughts weren't really fixed realities, but simply movements of the mind that was thinking them. Of course, as soon as I glimpsed that, the realization passed and I was back to comparing myself against the other students. But that brief moment of clarity was profound.

As my father explained after we'd finished, the point of the exercise was to recognize that there really is no difference between the mind that thinks and thoughts that come and go in the mind. The mind itself and the thoughts, emotions, and sensations that arise, abide, and disappear in the mind are equal expressions of emptiness—that is, the open-ended possibility for anything to occur. If the mind is not a "thing" but an event, then all the thoughts, feelings, and sensations that occur in what we think of as the mind are likewise events. As we begin to rest in the experience of mind and thoughts as inseparable, like two sides of the same coin, we begin to grasp the true meaning of clarity as an infinitely expansive state of awareness.

A lot of people think that meditation means achieving some unusually vivid state, completely unlike anything they've experienced before. They mentally squeeze themselves, thinking, *I've got to attain a higher level of consciousness. . . . I should be seeing something wonderful, like rainbow lights or images of pure realms. . . . I should be glowing in the dark.*

That's called *trying too hard,* and believe me, I've done it, as have a lot of other people I've come to know over the years.

Not long ago, I met someone who was causing problems for himself by trying too hard. I was sitting in the Delhi airport waiting to board a plane for Europe, when a man approached me and asked if I was a Buddhist monk. I replied that I was. He then asked me if I knew how to meditate, and when I replied that I did, he asked, "What's your experience like?"

"Fine," I answered.

"You don't find it difficult?"

"No," I said, "not really."

He shook his head and sighed. "Meditation is so hard for me," he explained. "After fifteen or twenty minutes, I start getting dizzy. And if I try to go on longer than that, I sometimes even throw up."

I told him that it sounded to me as though he was too tense, and that he should maybe try to relax more when he practiced.

"No," the man replied. "When I try to relax, I get even dizzier."

His problem seemed strange, and because he appeared genuinely interested in finding a solution, I asked him to sit across from me and meditate while I just watched him. After he settled in the seat opposite me, his arms, legs, and chest stiffened dramatically. His eyes bulged; a terrible grimace spread across his face; his eyebrows shot upward; and even his ears seemed to pull away from his head. His body was so tense he started shaking.

Just watching him, I thought I might get dizzy myself, so I said, "Okay, please stop."

He relaxed his muscles, the grimace vanished from his face, and his eyes, ears, and eyebrows returned to normal. Eagerly, he looked at me for advice.

"All right," I said. "Now I'm going to meditate, while you watch me the way I watched you."

I just sat in my seat as I normally do, with my spine straight, my muscles relaxed, my hands resting gently in my lap, and looking forward without any particular strain as I rested my mind with bare attention on the present moment. I watched the man looking at me from head to toe, toe to head, and head to toe again. Then I simply came out of meditation and told him that that was how I meditated.

After a moment he nodded slowly and said, "I think I get it."

Just then it was announced that our plane was ready for boarding. Since he and I were seated in different sections of the plane, we boarded separately and I didn't see him at all during the flight.

After we landed, I saw him again among the passengers disembarking from the plane. He waved, and as he approached me he said, "You know, I tried practicing the way you showed me, and through the whole flight I was able to meditate without getting dizzy. I think I finally understand what it means to relax in meditation. Thank you so much!"

It's certainly possible to have vivid experiences when you try too hard, but the more typical results can be grouped into three general types of experiences. The first is that the attempt to become aware of all the thoughts, feelings, and sensations rushing through your mind is simply exhausting, and as a result you may find your mind becoming tired or dull. The second is that the attempt to observe every thought, emotion, and sensation generates a sense of restlessness or agitation. The third is that you may discover that your mind goes completely blank: Every thought, emotion, feeling, or perception you observe passes so quickly that it simply eludes your awareness. In any of these cases, you might reasonably conclude that meditation isn't the great experience you imagined it might be.

Actually, the essence of meditation practice is to let go of all your expectations about meditation. All the qualities of your natural mind—peace, openness, relaxation, and clarity—are present in your mind just as it is. You don't have to do anything different. You don't have to shift or change your awareness. All you have to do while observing your mind is to recognize the qualities it already has.

LIGHTING UP THE DARK

> *You cannot separate a lit area and a shaded area from one another, they are so close.*
>
> —TULKU URGYEN RINPOCHE, *As It Is*, Volume 1,
> translated by Erik Pema Kunsang

Learning to appreciate the clarity of the mind is a gradual process, just like developing an awareness of emptiness. First you get the main point, slowly grow more familiar with it, and then just continue training in recognition. Some texts actually compare this slow course of recognition to an old cow peeing—a nice, down-to-earth description that keeps us from thinking of the process as something terribly difficult or abstract. Unless, however, you're a Tibetan nomad or happen to have been brought up on a farm, the comparison might not be immediately clear, so let me explain. An old cow doesn't pee in one quick burst, but in a slow, steady stream. It may not start out as much and it doesn't end quickly, either. In fact, the cow may walk several yards while in the process, continuing to graze. But when it's over—what a relief!

Like emptiness, the true nature of clarity is impossible to define completely without turning it into some sort of concept that you can tuck away in a mental pocket, thinking, *Okay, I get it, my mind is clear, now what?* Clarity in its pure form has to be experienced. And when you experience it, there's no "Now what?" You just get it.

If you think about the difficulty of trying to describe something that is essentially beyond description, you can probably understand something of the challenge the Buddha must have faced in trying to explain the nature of mind to his students—who were no doubt people just like ourselves, looking for clear-cut definitions that they could file away intellectually, making them feel momentarily proud that they were smarter and more sensitive than the rest of the world.

To avoid this trap, the Buddha, as we've seen, chose to describe the indescribable through metaphors and stories. In order to offer us a way to understand clarity in terms of everyday experience, he used the same analogy he used to describe emptiness, that of a dream.

He asked us to imagine the total darkness of sleep, with our eyes closed, the curtains shut, and our minds descending into a state of total blankness. Yet within this darkness, he explained, forms and experiences begin to appear. We encounter people—some familiar, others strangers. We may find ourselves in places we've known or places freshly imagined. The events we experience may be echoes of things we've experienced in waking life, or they may be completely new, never before imagined. In dreams, any and all experiences are possible, and the light that illumines and distinguishes the various people, places, and events within the darkness of sleep is an aspect of the pure clarity of mind.

The main difference between the dream example and true clarity is that even while dreaming, most of us still make a distinction between ourselves and others, and the places and events we experience. When we truly recognize clarity, we perceive no such distinction. Natural mind is indivisible. It's not as if *I'm* experiencing clarity over here, and *you're* experiencing clarity over there. Clarity, like emptiness, is infinite: It has no limits, no starting point and no end. The more deeply we examine our minds, the less possible it becomes to find a clear distinction between where our own mind ends and others' begin.

As this begins to happen, the sense of difference between "self" and "other" gives way to a gentler and more fluid sense of identification with other beings and with the world around us. And it's through this sense of identification that we start to recognize that the world may not be such a scary place after all: that enemies aren't enemies but people like ourselves, longing for happiness and seeking it the best way they know how, and that everyone possesses the insight, the wisdom, and the understanding to see past apparent differences and discover solutions that benefit not just ourselves but everyone around us.

APPEARANCE AND ILLUSION

> *Seeing the meaningful as meaningful, and the meaningless as meaningless, one is capable of genuine understanding.*
> —The Dhammapada,
> translated by Eknath Easwaran

The mind is like a stage magician, however. It can make us see things that aren't really there. Most of us are enthralled by the illusions our minds create, and we actually encourage ourselves to produce more and more outrageous fantasies. The sheer *drama* becomes addictive, producing what some of my students call an "adrenaline rush" or a "high" that makes us, or our problems, feel bigger than life—even when the situation that produces it is scary.

Just as we applaud the magician's trick of pulling a rabbit out of a hat, we watch horror movies, read suspense novels, get involved in difficult relationships, and fight with our bosses and coworkers. In a strange way—perhaps related to the most ancient, reptilian layer of the brain—we actually enjoy the tension these experiences provide. By strengthening our sense of "me" against "them," they confirm our sense of individuality—which, as we saw in the last chapter, is itself actually an appearance, lacking inherent reality.

Some cognitive psychologists I've spoken with have compared the human mind to a movie projector. Just as a projector casts images onto a screen, the mind projects sensory phenomena onto a type of cognitive screen—a context that we think of as the "external world"—while projecting thoughts, feelings, and sensations onto another type of screen or context that we refer to as our inner world, or "me."

That's getting close to a Buddhist perspective on absolute and relative reality. *Absolute reality* is emptiness, a condition in which perceptions are intuitively recognized as an infinite and transitory flow of possible experiences. When you begin to recognize perceptions as nothing more than fleeting, circumstantial events, they don't weigh as heavily on you, and the whole dualistic structure of "self" and "other" begins to soften. *Relative reality* is the sum of experiences arising from the mistaken idea that whatever you perceive is real in and of itself.

The habit of thinking that things exist "out there" in the world or "in here" is hard to give up, though. It means letting go of all the illusions you cherish, and recognizing that everything you project, everything you think of as "other," is in fact a spontaneous expression of your own mind. It means letting go of *ideas about reality* and instead experiencing the flow of reality *as it is*. At the same time, you don't have to completely disengage from your perceptions. You don't have to isolate yourself in a cave or mountain retreat. You can *enjoy* your perceptions without actively engaging them, looking at them in the same way you'd look at the objects you'd experience in a dream. You can actually begin to marvel at the variety of experiences that present themselves to you.

Through recognizing the distinction between appearance and illusion, you give yourself permission to acknowledge that some of your perceptions might be wrong or biased, that your ideas of how things ought to be may have solidified to the degree that you can't see any other point of view but your own. When I began to recognize the emptiness and clarity of my own mind, my life became richer and more vivid in ways I never could have imagined. Once I shed my ideas about how things should be, I became free to respond to my experience exactly as it was and exactly as I was, right there, right then.

THE UNION OF CLARITY AND EMPTINESS

Our true nature has inexhaustible properties.

—MAITREYA, *The Mahayana Uttaratantra Shastra,*
translated by Rosemarie Fuchs

It's said that the Buddha taught 84,000 methods to help people at various levels of understanding recognize the power of the mind. I haven't studied them all, so I can't swear that the number is exact. He may have taught 83,999 or 84,001. The essence of his teachings, however, can be reduced to a single point: *The mind is the source of all experience, and by changing the direction of the mind, we can change the quality of everything we experience.* When you transform your mind, everything you experience is transformed. It's like putting on a pair of yellow glasses: Suddenly, everything you see is yellow. If you put on a pair of green glasses, everything you see is green.

Clarity, in this sense, may be understood as the creative aspect of mind. Everything you perceive, you perceive through the power of your awareness. There are truly no limits to the creative ability of your mind. This creative aspect is the natural consequence of the union of emptiness and clarity. It's known in Tibetan as *magakpa,* or "unimpededness." Sometimes *magakpa* is translated as "power" or "ability," but the meaning is the same: the freedom of the mind to experience anything and everything whatsoever.

To the extent that you can acknowledge the true power of your mind, you can begin to exercise more control over your experience. Pain, sadness, fear, anxiety, and all other forms of suffering no longer disrupt your life as forcefully as they used to. Experiences that once seemed to be obstacles become opportunities for deepening your understanding of the mind's unimpeded nature.

Everyone experiences sensations of pain and pleasure throughout their lives. Most of these sensations appear to have some sort of a physical basis. Having a massage, eating good food, or taking a warm bath would generally be considered physically pleasant experiences. Burning a finger, getting an injection, or being stuck in traffic on a hot day in a car without air-conditioning would be considered physically unpleasant. Actually, though, whether you experience these things as painful or pleasurable doesn't depend on the physical sensations in themselves, but on your perception of them.

For example, some people can't stand feeling hot or cold. They say they'll die if they have to go outside in hot weather. Even a few drops of sweat can make them feel extremely uncomfortable. In winter, they can't bear even a few flakes of snow on their heads. But if a doctor they trust tells them that spending ten minutes every day in a sauna will improve their physical condition, they'll often follow the advice, seeking out and even paying for an experience they previously couldn't stand. They'll sit in the sauna thinking, *How nice, I'm sweating! This is really good!* They do this because they've allowed themselves to shift their mental perception about being hot and sweaty. Heat and sweat are just phenomena to which they've assigned different meanings. And if the doctor further tells them that a cold shower after the sauna will improve their circulation, they learn to accept the cold, and even come to consider it refreshing.

Psychologists often refer to this sort of transformation as "cognitive restructuring." Through applying *intention* as well as *attention* to an experience, a person is able to shift the meaning of an experience from a painful or intolerable context to one that is tolerable or pleasant. Over time, cognitive restructuring establishes new neuronal pathways in the brain, particularly in the limbic region, where most sensations of pain and pleasure are recognized and processed.

If our perceptions really are mental constructs conditioned by past experiences and present expectations, then what we focus on and how we focus become important factors in determining our experience. And the more deeply we believe something is true, the more likely it will become true in terms of our experience. So if we believe we're weak, stupid, or incompetent, then no matter what our real qualities are, and no matter how differently our friends and associates see us, we'll experience ourselves as weak, stupid, or incompetent.

What happens when you begin to recognize your experiences as your own projections? What happens when you begin to lose your fear of the people around you and conditions you used to dread? Well, from one point of view—nothing. From another point of view—everything.

7

COMPASSION: SURVIVAL OF THE KINDEST

Immense compassion springs forth spontaneously toward all sentient beings who suffer as prisoners of their illusions.
—Kalu Rinpoche, *Luminous Mind: The Way of the Buddha,*
translated by Maria Montenegro

IMAGINE SPENDING YOUR life in a little room with only one locked window so dirty it barely admits any light. You'd probably think the world was a pretty dim and dreary place, full of strangely shaped creatures that cast terrifying shadows against the dirty glass as they passed your room. But suppose one day you spill some water on the window, or a bit of rain dribbles in after a storm, and you use a rag or a corner of your shirtsleeve to dry it off. And as you do that, a little of the dirt that had accumulated on the glass comes away. Suddenly a small patch of light comes through the glass. Curious, you might rub a little harder, and as more dirt comes away, more light streams in. *Maybe,* you think, *the world isn't so dark and dreary after all. Maybe it's the window.*

You go to the sink and get more water (and maybe a few more rags), and rub and rub until the whole surface of the window is free of dirt and grime. The light simply pours in, and you recognize, perhaps for the first time, that all those strangely shaped shadows that used to scare you every time they passed are people—just like you! And from the depths of your awareness arises an instinctive urge to form a social bond—to go out there on the street and just be with them.

In truth, you haven't changed anything at all. The world, the light, and the people were always there. You just couldn't see them because your vision was obscured. But now you see it all, and what a difference it makes!

This is what, in the Buddhist tradition, we call the dawning of compassion, the awakening of an inborn capacity to identify with and understand the experience of others.

THE BIOLOGY OF COMPASSION

Those with great compassion possess all the Buddha's teaching.
> —*The Sutra That Completely Encapsulates the Dharma,*
> translated by the Padmakara Translation Group

The Buddhist understanding of compassion is, in some ways, a bit different from the ordinary sense of the word. For Buddhists, compassion doesn't simply mean feeling sorry for other people. The Tibetan term—*nying-jay*—implies an utterly direct expansion of the heart. Probably the closest English translation of *nying-jay* is "love"—but a type of love without attachment or any expectation of getting anything in return. Compassion, in Tibetan terms, is a spontaneous feeling of connection with all living things. What you feel, I feel; what I feel, you feel. There's no difference between us.

Biologically, we're programmed to respond to our environment in fairly simple terms of avoiding threats to our survival and grasping for opportunities to enhance our well-being. We only need to flip through the pages of a history book to see that the story of human development is very frequently a tale of violence written in the blood of weaker beings.

Yet it seems that the same biological programming that drives us toward violence and cruelty also provides us with emotions that not only inhibit aggression but also move us to act in ways that override the impulse for personal survival in the service of others. I was struck by a remark made by Harvard professor Jerome Kagan during his presentation at the 2003 Mind and Life Institute conference, when he noted that along with our tendency toward aggression, our survival instinct has provided us with "an even stronger biological bias for kindness, compassion, love, and nurture."[1]

I have been told many stories about the number of people who risked their lives during the Second World War to give refuge to European Jews hunted by the Nazis, and of the unnamed heroes of the present day who sacrifice their own welfare to help the victims of war, famine, and tyranny in countries around the world. In addition, many of my Western students are parents who sacrifice an enormous amount of time and energy shuttling their children between sports competitions, musical activities, and other events, while patiently putting money aside for their children's education.

Such sacrifices do seem, on an individual level, to indicate a set of biological factors that transcend personal fears and desires. The simple fact that we've been able to build societies and civilizations that at least acknowledge the need to protect and care for the poor, the weak, and the defenseless supports Professor Kagan's conclusion that "an ethical sense is a biological feature of our species."[2]

His remarks resonate almost completely with the essence of the Buddha's teachings: The more clearly we see things as they are, the more willing and able we become to open our hearts toward other beings. When we recognize that others experience pain and unhappiness because they don't recognize their real nature, we're spontaneously moved by a profound wish for them to experience the same sense of peace and clarity that we've begun to know.

THE AGREEMENT TO DISAGREE

> *Hot seeds will produce hot fruits.*
> *Sweet seeds will produce sweet fruits.*
>
> —*The Questions of Surata Sutra,*
> translated by Elizabeth M. Callahan

From what I've learned, most conflicts between people stem from a misunderstanding of one another's motives. We all have our reasons for doing what we do and saying what we say. The more we allow ourselves to be guided by compassion—to pause for a moment and try to see where another person is coming from—the less likely we are to engage in conflict. And even when problems do arise, if we take a deep breath and listen with an open heart, we'll find ourselves able to handle the conflict more effectively—to calm the waters, so to speak, and resolve our differences in such a manner that everyone is satisfied, and no one ends up as the "winner" or the "loser."

For example, I have a Tibetan friend in India who lived next door to a man who had a bad-tempered dog. In India, unlike in other countries, the walls surrounding the front yard of a house are very tall, with doors instead of gates. The entrances to my friend's yard and his neighbor's yard were very close, and every time my friend came out of his door, the dog would tear out of his neighbor's door, barking, growling, fur bristling—an altogether scary experience for my friend. As if that weren't bad enough, the dog had also developed a habit of pushing through the door into my friend's yard, again barking and snarling, making a terrible disturbance.

My friend spent a long time considering how to punish the dog for its bad behavior. At last he hit on the idea of propping open the door to his front yard just a bit, and loosely piling a few small, heavy objects on top of it. The next time the dog pushed open the door, the objects would fall, teaching him a painful lesson he would never forget.

After setting his trap one Saturday morning , my friend sat by his front window, watching and waiting for the dog to enter the yard. Time passed and the dog never came. After a while my friend set out his daily prayer texts and started chanting, glancing up from his texts every once in a while to look out the window into the yard. Still, the dog failed to appear. At a certain point in his chanting, my friend came to a very ancient prayer of aspiration known as "The Four Immeasurables," which begins with the following lines:

> *May all sentient beings have happiness and the causes of*
> * happiness.*
> *May all sentient beings be free from suffering and the causes*
> * of suffering.*

In the middle of chanting this prayer, it suddenly occurred to him that the dog was a sentient being and that in having deliberately set a trap, he would cause the dog pain and suffering. *If I chant this,* he thought, *I'll be lying. Maybe I should stop chanting.*

But that didn't feel right, since the Four Immeasurables prayer was part of his daily practice. He started the prayer again, making an earnest effort to develop compassion toward dogs, but halfway through he stopped himself, thinking, *No! That dog is very bad. He causes me a lot of harm. I don't want him to be free of suffering or to achieve happiness.*

He thought about this problem for a while, until a solution finally came to him. He could change one small word of the prayer. And he began to chant:

> *May SOME sentient beings have happiness and the causes of happiness.*
> *May SOME sentient beings be free from suffering and the causes of suffering.*

He felt quite happy with his solution. After he'd finished his prayers, eaten his lunch, and forgotten about the dog, he decided to go out for a walk before the day was over. In his haste, he forgot about the trap he'd set, and as soon as he pulled open the door to his yard, all the heavy things he'd piled up on its edge fell on his head.

It was, to say the least, a rude awakening.

Yet, as a result of his pain, my friend realized something of great importance. By excluding any beings from the possibility of achieving happiness and freedom from suffering, he had also excluded himself. Recognizing that he himself was the victim of his own lack of compassion, he decided to change his tactics.

The next day, when he went out for his morning walk, my friend carried with him a small piece of *tsampa*—a kind of dough made of ground barley, salt, tea, and lumps of butter—that Tibetans usually eat for breakfast. As soon as he stepped out his door, the neighbor's dog came rushing out, barking and snarling as usual; but instead of cursing the dog, my friend simply threw him the piece of the tsampa he was carrying. Completely surprised in mid-bark, the dog caught the tsampa in his mouth and began to chew—still bristling and growling, but distracted from his attack by the offering of food.

This little game continued over the next several days. My friend would step out of his yard, the dog would come running out, and in mid-bark would catch the bit of tsampa my friend threw him. After a few days my friend noticed that even though it kept growling while chewing on the tsampa, the dog had started to wag its tail. By the end of a week, the dog was no longer bounding out ready to attack, but instead ran out to greet my friend, happily expecting a treat. Eventually the relationship between the two developed to the point where the dog would come trotting quietly into my friend's yard to sit with him in the sun, while he recited his daily prayers—quite contentedly now able to pray for the happiness and freedom of *all* sentient beings.

Once we recognize that other sentient beings—people, animals, and even insects—are just like us, that their basic motivation is to experience peace and to avoid suffering, then, when someone acts in some way or says something that is against our wishes, we're able to have some basis for understanding: "Oh, well, this person (or whatever) is coming from this position because, just like me, they want to be happy and they want to avoid suffering. That's their basic purpose. They're not out to get me; they're only doing what they think they need to do."

Compassion is the spontaneous wisdom of the heart. It's always with us. It always has been, and always will be. When it arises in us, we've simply learned to see how strong and safe we really are.

8

WHY ARE WE UNHAPPY?

All sentient beings tend to act in a way that is unbeneficial.
—JAMGÖN KONGTRUL, *The Torch of Certainty,*
translated by Judith Hanson

AFTER ALMOST TEN years of teaching in more than twenty countries around the world, I've seen a lot of strange and wonderful things, and heard a lot of strange and wonderful stories from people who have spoken up at public teachings or come to me for private counseling. What's surprised me most, though, was to see that people living in places where material comforts were widely available appeared to experience a depth of suffering similar to what I'd seen among those who lived in places that weren't quite so materially developed. The expression of suffering I witnessed was different in some respects from what I'd become accustomed to witnessing in India and Nepal, but its strength was palpable.

I began to sense this level of unhappiness during my first few visits to the West, when my hosts would take me to see the great landmarks in their cities. When I first saw places like the Empire State Building or the Eiffel Tower, I couldn't help but be struck by the genius of the designers and the degree of cooperation and determination that must have been required of the people who built these structures. But when we arrived at the observation deck, I'd find the view blocked by barbed-wire fencing and the whole area patrolled by guards. When I asked my hosts about the guards and fences, they explained that the precautions were necessary to keep people from killing themselves by

jumping from the heights. It seemed immeasurably sad to me that societies capable of producing such wonders would need to impose
strict measures to keep people from using these beautiful monuments
as platforms for suicide.

The security measures didn't detract at all from my appreciation
of the beauty of these places or the technological skill required to
build them. But after I'd visited a few of these places, the security precautions began to "click into place" with something else I'd begun to
notice. Although people living in materially comfortable cultures
tended to smile easily enough, their eyes almost always betrayed a
sense of dissatisfaction and even desperation. And the questions people asked during both public and private talks often seemed to revolve
around how to become better or stronger than they were, or how to get
over "self-hatred."

The more widely I traveled, the clearer it became to me that people
living in societies characterized by technological and material achievements were just as likely to feel pain, anxiety, loneliness, isolation, and
despair as people who lived in comparatively less-developed areas.
After a few years of asking some very pointed questions in public
teachings and in private counseling sessions, I began to see that when
the pace of external or material progress exceeded the development of
inner knowledge, people seemed to suffer deep emotional conflicts
without any internal method of dealing with them. An abundance of
material items provides such a variety of external distractions that people lose the connection to their inner lives.

Just think, for example, about the number of people who desperately look for a sense of excitement by going to a new restaurant, starting a new relationship, or moving to a different job. For a while the
newness does seem to provide some sense of stimulation. But eventually the excitement dies down; the new sensations, new friends, or
new responsibilities become commonplace. Whatever happiness they
originally felt dissolves.

So they try a new strategy, like going to the beach. And for a while
that seems satisfying, too. The sun feels warm, the water feels great,
and there's a whole new crowd of people to meet, and maybe new and
exciting activities to try, like jet skiing or parasailing. But after a while,

even the beach gets boring. The same old conversations are repeated over and over, the sand feels gritty on your skin, the sun is too strong or hides behind clouds, and the ocean gets cold. So it's time to move on, try a different beach, maybe in a different country. The mind produces its own sort of mantra: "I want to go to Tahiti . . . Tahiti . . . Tahiti. . . ."

The trouble with all of these solutions is that they are, by nature, temporary. All phenomena are the results of the coming together of causes and conditions, and therefore inevitably undergo some type of change. When the underlying causes that produced and perpetuated an experience of happiness change, most people end up blaming either external conditions (other people, a place, the weather, etc.) or themselves ("I should have said something nicer or smarter," "I should have gone somewhere else"). However, because it reflects a loss of confidence in oneself, or in the things we're taught to believe *should* bring us happiness, blame only makes the search for happiness more difficult.

The more problematic issue is that most people don't have a very clear idea of what happiness is, and consequently find themselves creating conditions that lead them back to the dissatisfaction they so desperately seek to eliminate. That being the case, it would be a good idea to look at happiness, unhappiness, and their underlying causes a bit more closely.

THE EMOTIONAL BODY

> *There is no one center for emotion, just as there is none for playing tennis.*
>
> —RICHARD DAVIDSON, quoted in Daniel Goleman,
> *Destructive Emotions: How Can We Overcome Them?*

Our bodies play a much bigger role in the generation of emotions than most of us recognize. The process begins with perception—which we already know involves the passing of information from sensory organs to the brain, where a conceptual representation of an object is created. Most of us would assume, quite naturally, that once the object is perceived and recognized, an emotional response is produced, which in turn generates some sort of physical reaction.

In fact, the opposite occurs. At the same time that the thalamus sends its messages higher up to the analytical regions of the brain, it sends a simultaneous "red alert" message to the amygdala, the funny little walnut-shaped neuronal structure in the limbic region, which, as described earlier, governs emotional responses, particularly fear and anger. Because the thalamus and the amygdala are very close to each other, this red alert signal travels much more quickly than the messages sent to the neocortex. Upon receiving it, the amygdala immediately sets in motion a series of physical responses that activate the heart; the lungs; major muscles groups in the arms, chest, abdomen, and legs; and the organs responsible for producing hormones like adrenaline. Only *after* the body responds does the analytical part of the brain interpret the physical reactions in terms of a specific emotion. In other words, you don't see something scary, feel fear, and then run. You see something scary, start to run (as your heart pounds and adrenaline surges through your body), and *then* interpret the body's reaction as fear. In most cases, though, once the rest of your brain has caught up with your body, which takes only a few milliseconds, you're able to assess your reactions and determine whether they're appropriate, as well as adjust your behavior to fit the particular situation.

The results of this assessment can actually be measured by technology that has only recently become available to scientists. Emotions such as fear, disgust, and loathing appear in part as a heightened activation of neurons in the right frontal lobe, the region of the neocortex located at the very front of the right side of the brain. Meanwhile, emotions such as joy, love, compassion, and confidence can be measured in terms of relatively greater activity among the neurons in the left frontal lobe.

In some instances, I've been told, our ability to evaluate our reactions is inhibited, and we find ourselves responding to a situation without thinking. In such cases the amygdala's response is so strong that it short-circuits the reaction of the higher brain structures. Such a powerful "emergency response" mechanism undoubtedly has important survival benefits, enabling us to immediately recognize foods that once made us sick, or to avoid aggressive animals. But because the neuronal patterns stored in the amygdala can be triggered easily by events that bear even a slight resemblance to an earlier incident, they can distort our perception of present-moment events.

STATES AND TRAITS

Everything depends on circumstance.

—PATRUL RINPOCHE, *The Words of My Perfect Teacher,*
translated by the Padmakara Translation Group

From a scientific perspective, emotions are viewed in terms of short-term events and longer-lasting conditions. Short-term emotions might include the sudden burst of anger we experience when we're fixing something around the house and accidentally hit our thumb with a hammer, or the swell of pride we feel when someone pays us a genuine compliment. In scientific terms, these relatively short-term events are often referred to as *states.*

Emotions that continue over time and across a variety of situations, such as the love someone feels for a child or a lingering resentment over something that happened in the past, are referred to as *traits* or *temperamental qualities,* which most of us regard as indicators of a person's character. For example, we tend to say that a person who is usually smiling and energetic and always has nice things to say to other people is a "cheerful" person, while we tend to think of someone who frowns a lot, runs around in a hurry, hunches over his desk, and loses his temper over little things as a "tense" person.

The difference between states and traits is fairly obvious, even to someone who doesn't have a science degree. If you hit your thumb with a hammer, chances are good that the anger you experience will pass fairly quickly and won't cause you to be afraid of hammers for the rest of your life. Emotional traits are more subtle. In most cases we're able to recognize whether we wake up anxious or excited day after day, while indications of our temperament gradually become evident over time to others with whom we're in close contact.

Emotional states are fairly quick bursts of neuronal gossip. Traits, on the other hand, are more like the neuronal equivalent of committed relationships. The origins of these long-lasting connections may vary.

Some may have a genetic basis, others may be caused by serious trauma, and still others may simply have developed as a result of sustained or repeated experiences—the life training we receive as children and young adults.

Whatever their origin, emotional traits have a conditioning effect on the way people characterize and respond to their everyday experiences. Someone predisposed to fear or depression, for example, is more likely to approach situations with a sense of trepidation, whereas someone disposed toward confidence will approach the same situation with much more poise and assurance.

CONDITIONING FACTORS

Suffering follows a negative thought as the wheels of a cart follow the oxen that draw it.

—*The Dhammapada,*
translated by Eknath Easwaran

Biology and neuroscience tell us what's going on in our brains when we experience pleasant or unpleasant emotions. Buddhism helps us not only to describe such experiences more explicitly to ourselves, but also provides us with the means to go about changing our thoughts, feelings, and perceptions so that on a basic, cellular level we can become happier, more peaceful, and more loving human beings.

Whether looked at subjectively through mindful observation taught by the Buddha, or objectively, through the technology available in modern laboratories, what we call the mind emerges as a constantly shifting collision of two basic events: bare recognition (the simple awareness that something is happening) and conditioning factors (the processes that not only describe what we perceive, but also determine our responses). *All* mental activity, in other words, evolves from the combined activity of bare perception and long-term neuronal associations.

One of the lessons repeated again and again by my teacher Saljay Rinpoche was that if I wanted to be happy, I had to learn to recognize and work with the conditioning factors that produce compulsive or trait-bound reactions. The essence of his teaching was that any factor can be understood as compulsive to the degree that it obscures our ability to see things as they are, without judgment. If someone is yelling at us, for example, we rarely take the time to distinguish between the bare recognition "Oh, this person is raising his voice and saying such and such words" and the emotional response "This person is a jerk." Instead, we tend to combine bare perception and our emotional response into a single package: "This person is screaming at me *because* he's a jerk."

But if we could step back to look at the situation more objectively, we might see that people who yell at us are upset over something that may have nothing to do with us. Maybe they just got criticized by someone higher up and are afraid of getting fired. Maybe they just found out that someone close to them is very sick. Or maybe they had an argument with a friend or a partner and didn't sleep well afterward. Sadly, the influence of conditioning is so strong that we rarely remember that we *can* step back. And because our understanding is limited, we mistake the little part we do see for the whole truth.

How can we respond appropriately when our vision is so limited, when we don't have all the facts? If we apply the standard of American courts to tell "the whole truth and nothing but the truth" about our everyday experience, we must recognize that the "whole truth" is that everyone just wants to be happy. The truly sad thing is that most people seek happiness in ways that actually sabotage their attempts. If we could see the whole truth of any situation, our only response would be one of compassion.

MENTAL AFFLICTIONS

> *By whom and how were the weapons of hell created?*
> —Śāntideva, *The Bodhicaryāvatāra*,
> translated by Kate Crosby and Andrew Skilton

The conditioning factors are often referred to in Buddhist terms as "mental afflictions," or sometimes "poisons." Although the texts of Buddhist psychology examine a wide range of conditioning factors, all of them agree in identifying three primary afflictions that form the basis of all other factors that inhibit our ability to see things as they really are: *ignorance, attachment,* and *aversion.*

Ignorance

Ignorance is a fundamental inability to recognize the infinite potential, clarity, and power of our own minds, as if we were looking at the world through colored glass: Whatever we see is disguised or distorted by the colors of the glass. On the most essential level, ignorance distorts the basically open experience of awareness into dualistic distinctions between inherently existing categories of "self" and "other."

Ignorance is thus a twofold problem. Once we establish the neuronal habit of identifying ourselves as a single, independently existing "self," we inevitably start to see whatever is not "self" as "other." "Other" can be anything: a table, a banana, another person, or even something this "self" is thinking or feeling. Everything we experience becomes, in a sense, a stranger. And as we become accustomed to distinguishing between "self" and "other," we lock ourselves into a dualistic mode of perception, drawing conceptual boundaries between our "self" and the rest of the world "out there," a world that seems so vast that we almost can't help but begin to think of ourselves as very small, limited, and vulnerable. We begin looking at other people, material objects, and so on as potential sources of happiness and unhappiness, and life becomes a struggle to get what we need in order to be happy before somebody else grabs it.

This struggle is known in Sanskrit as *samsara,* which literally means "wheel" or "circle." Specifically, samsara refers to the wheel or circle of unhappiness, a habit of running around in circles, chasing after the same experiences again and again, each time expecting a different result. If you've ever watched a dog or a cat chasing its own tail, you've seen the essence of samsara. And even though it might be funny to watch an animal chase its tail, it's not so funny when your own mind does the same thing.

The opposite of samsara is *nirvana,* a term that is almost as completely misunderstood as *emptiness.* A Sanskrit word roughly translated as "extinguishing" or "blowing out" (as in the blowing out of the flame of a candle), nirvana is often interpreted as a state of total bliss or happiness, arising from the extinguishing or "blowing out" of the ego or the idea of "self." This interpretation is accurate to a certain extent, except that it doesn't take into account that most of us live as embodied beings going about our lives in the relatively real world of moral, ethical, legal, and physical distinctions.

Trying to live in this world without abiding by its relative distinctions would be as foolish and difficult as trying to avoid the consequences of being born right- or left-handed. What would be the point? A more precise interpretation of nirvana is the adoption of a broad perspective that admits all experiences, pleasurable or painful, as aspects of awareness. Naturally, most people would prefer to experience only the "high notes" of happiness. But as a student of mine recently pointed out, eliminating the "low notes" from a Beethoven symphony—or any modern song, for that matter—would result in a pretty cheap and tinny experience.

Samsara and nirvana are perhaps best understood as points of view. Samsara is a point of view based primarily on defining and identifying with experiences as either painful or unpleasant. Nirvana is a fundamentally objective state of mind: an acceptance of experience without judgments, which opens us to the potential for seeing solutions that may not be directly connected to our survival as individuals, but rather to the survival of all sentient beings.

Which brings us to the second of the three primary mental afflictions.

Attachment

The perception of "self" as separate from "others" is, as discussed earlier, an essentially biological mechanism—an established pattern of neuronal gossip that consistently signals other parts of the nervous system that each of us is a distinct, independently existing creature that needs certain things in order to perpetuate its existence. Because we live in physical bodies, some of these things we need, such as oxygen, food, and water, are truly indispensable. In addition, studies of infant survival that people have discussed with me have shown that survival requires a certain amount of physical nurturing.[1] We need to be touched; we need to be spoken to; we need the simple fact of our existence to be acknowledged.

Problems begin, however, when we generalize biologically essential things into areas that have nothing to do with basic survival. In Buddhist terms, this generalization is known as "attachment" or "desire"—which, like ignorance, can be seen as having a purely neurological basis.

When we experience something like chocolate, for example, as pleasant, we establish a neuronal connection that equates chocolate with the physical sensation of enjoyment. This is not to say that chocolate in itself is a good or bad thing. There are lots of chemicals in chocolate that create a physical sensation of pleasure. It's our neuronal attachment to chocolate that creates problems.

Attachment is in many ways comparable to addiction, a compulsive dependency on external objects or experiences to manufacture an illusion of wholeness. Unfortunately, like other addictions, attachment becomes more intense over time. Whatever satisfaction we might experience when we attain something or someone we desire doesn't last. Whatever or whoever made us happy today, this month, or this year is bound to change. Change is the only constant of relative reality.

The Buddha compared attachment to drinking salt water from an ocean. The more we drink, the thirstier we get. Likewise, when our mind is conditioned by attachment, however much we have, we never really experience contentment. We lose the ability to distinguish between the bare experience of happiness and whatever objects temporarily make us happy. As a result, we not only become dependent on the object, but we also reinforce the neuronal patterns that condition us to rely on an external source to give us happiness.

You can substitute any number of objects for chocolate. For some people, relationships are the key to happiness. When they see someone they think is attractive, they try all kinds of ways to approach him or her. But if they finally manage to become involved with that person, the relationship doesn't turn out to be as satisfying as they imagined. Why? Because the object of their attachment is not really an external thing. It's a story spun by the neurons in the brain; and that story unfolds on many different levels, ranging from what they think they might gain from achieving what they desire to what they fear if they fail to get it.

Other people think they'd be really happy if they experienced an extreme stroke of good luck, like winning the lottery. But an interesting study by Philip Brinkman[2] that I heard about from one of my students has shown that people who had recently won a lottery were not that much happier than a control group who hadn't experienced the excitement of suddenly becoming rich. In fact, after the initial thrill wore off, the people who'd won a lottery reported finding *less* enjoyment in the everyday pleasures, like chatting with friends, getting compliments, or simply reading a magazine, than people who hadn't experienced such a major change.

The study reminded me of a story I heard not long ago about an old man who'd bought a ticket for a lottery worth more than a hundred million dollars. A short time after buying the ticket, he developed a heart problem and was sent to the hospital under the care of a doctor who ordered strict bed rest and absolutely forbade anything that would cause undue excitement. While the old man was in the hospital, his ticket actually won the lottery. Since he was in a hospital, of course, the old man didn't know about his good fortune, but his children and his wife found out and went to the hospital to tell the man the news.

On the way to his hospital room, they met his doctor and told him all about the old man's good fortune. As soon as they'd finished, the doctor pleaded with them not to say anything just yet. "He might get so excited," the doctor explained, "that he could die from the strain on his heart." The man's wife and children argued with the doctor, believing that the good news would help improve his condition. But in the end they agreed to let the doctor break the news, gently and slowly so as not to cause the man undue excitement.

While the man's wife and children sat waiting in the hall, the doctor went into his patient's room. He began by asking the man all sorts of questions about his symptoms, how he was feeling, and so on; and after a while, he asked, very casually, "Have you ever bought a ticket for the lottery?"

The old man replied that, in fact, he had bought a ticket just before coming to the hospital.

"If you won the lottery," the doctor asked, "how would you feel?"

"Well, if I do, that would be nice. If I don't, that would be fine, too. I'm an old man and won't live much longer. Whether I win or not, it doesn't really matter."

"You couldn't really feel that way," the doctor said, in the manner of someone speaking purely theoretically. "If you won, you'd be really excited, right?"

But the old man replied, "Not really. In fact, I'd be happy to give you half of it if you could find a way to make me feel better."

The doctor laughed. "Don't even think about it," he said. "I was just asking."

But the patient insisted, "No, I mean it. If I won the lottery, I really would give you half of what I won if you could make me feel better."

Again, the doctor laughed. "Why don't you write a letter," he joked, "saying you'd give me half?"

"Sure, why not?" the old man agreed, reaching over to the table next to his bed and picking up a pad of paper. Slowly, feebly, he wrote out a letter agreeing to give the doctor half of any lottery money he might win, signed it, and handed it to the doctor. When the doctor looked at the letter and the signature, he got so excited over the idea of getting so much money that he fell over dead on the spot.

As soon as the doctor fell, the old man started shouting. Hearing the noise, the man's wife and children feared that the doctor had been right all along, that the news really had been too exciting, and the old man had died from the strain on his heart. They rushed into the room, only to find the old man sitting up in his bed and the doctor crumpled on the floor. While the nurses and other hospital staff rushed around trying to revive the doctor, the old man's family quietly told him that he had won the lottery. Much to their surprise, he didn't seem all that excited about learning that he'd just won millions of dollars, and the news didn't do him any damage at all. In fact, after a few weeks his condition improved and he was released from the hospital. Certainly he was glad to enjoy his new wealth, but he wasn't all that attached to it. The doctor, on the other hand, had been so attached to the idea of having so much money, and his excitement was so great, that his heart couldn't bear the strain and he died.

Aversion

Every strong attachment generates an equally powerful fear that we'll either fail to get what we want or lose whatever we've already gained. This fear, in the language of Buddhism, is known as *aversion*: a resistance to the inevitable changes that occur as a consequence of the impermanent nature of relative reality.

The notion of a lasting, independently existing self urges us to expend enormous effort in resisting the inevitability of change, making sure that this "self" remains safe and secure. When we've achieved some condition that makes us feel whole and complete, we want everything to stay exactly as it is. The deeper our attachment to whatever provides us with this sense of completeness, the greater our fear of losing it, and the more brutal our pain if we do lose it.

In many ways, aversion is a self-fulfilling prophecy, compelling us to act in ways that practically guarantee that our efforts to attain whatever we think will bring us lasting peace, stability, and contentment will fail. Just think for a moment about how you act around someone to whom you feel a strong attraction. Do you behave like the suave, sophisticated, and self-confident person you'd like the other person to see, or do you suddenly become a tongue-tied goon? If this person talks and laughs with someone else, do you feel hurt or jealous, and betray your pain and jealousy in small or obvious ways? Do you become so fiercely attached to the other person to such a degree that he or she senses your desperation and begins to avoid you?

Aversion reinforces neuronal patterns that generate a mental construct of yourself as limited, weak, and incomplete. Because anything that might undermine the independence of this mentally constructed "self" is perceived as a threat, you unconsciously expend an enormous amount of energy on the lookout for potential dangers. Adrenaline rips through your body, your heart races, your muscles tense, and your lungs pump like mad. All these sensations are symptoms of stress, which, as I've heard from many scientists, can cause a huge variety of problems, including depression, sleeping disorders, digestive problems, rashes, thyroid and kidney malfunctions, high blood pressure, and even high cholesterol.

On a purely emotional level, aversion tends to manifest as anger and even hatred. Instead of recognizing that whatever unhappiness you feel is based on a mentally constructed image, you find it only "natural" to blame other people, external objects, or situations for your pain. When people behave in a way that appears to prevent you from obtaining what you desire, you begin to think of them as untrustworthy or mean, and you'll go out of your way either to avoid them or strike back at them. In the grip of anger, you see everyone and everything as enemies. As a result, your inner and outer worlds become smaller and smaller. You lose faith in yourself, and reinforce specific neuronal patterns that generate feelings of fear and vulnerability.

AFFLICTION OR OPPORTUNITY?

> *Consider the advantages of this rare human existence.*
> —JAMGÖN KONGTRUL, *The Torch of Certainty*,
> translated by Judith Hanson

It's easy to think of mental afflictions as defects of character. But that would be a devaluation of ourselves. Our capacity for emotions, for distinguishing between pain and pleasure, and for experiencing "gut responses" has played and continues to play a critical survival function, enabling us almost instantaneously to adapt to subtle changes in the world around us, and to formulate those adaptations consciously so that we can recall them at will and pass them along to succeeding generations.

Such extraordinary sensitivity reinforces one of the most basic lessons taught by the Buddha, which was to consider how precious this human life is, with all its freedoms and opportunities; how difficult it is to obtain such a life; and how easy it is to lose it.

It doesn't matter whether you believe that human life is a cosmic accident, a karmic lesson, or the work of a divine creator. If you simply pause to consider the huge variety and number of creatures that share the planet with us, compared with the relatively small percentage of human beings, you have to conclude that the chances of being born as a human being are extremely rare. And in demonstrating the extraordinary complexity and sensitivity of the human brain, modern science reminds us how fortunate we are to have been born human, with the very human capacity to feel and to sense the feelings of those around us.

From a Buddhist standpoint, the automatic nature of human

emotional tendencies represents an interesting challenge. It doesn't require a microscope to observe psychological habits; most people don't have to look any further than their last relationship. They begin by thinking, *This time it's going to be different.* A few weeks, months, or years later, they smack their heads, thinking, *Oh no, this is exactly the same type of relationship I was involved in before.*

Or you can look at your professional life. You start a new job thinking, *This time I'm not going to end up spending hours and hours working late, only to get criticized for not doing enough.* Yet three or four months into the job, you find yourself canceling appointments or calling friends to say, "I can't make dinner tonight. I have too much work to do."

Despite your best intentions, you find yourself repeating the same patterns while expecting a different result. Many of the people I've worked with over the years have talked about how they daydreamed about getting through the week so they could enjoy the weekend. But when the weekend is over, they're back at their desks for another week, daydreaming about the next weekend. Or they tell me about how they've invested enormous time and effort in completing a project, but never allow themselves to experience any sense of accomplishment because they have to start working on the next task on their list. Even when they're relaxing, they say they're preoccupied by something that happened the previous week, the previous month, or even the previous year, replaying scenes over and over in their minds, trying to figure out what they could have done to make the outcome more satisfying.

Fortunately, the more familiar we become with examining our minds, the closer we come to finding a solution to whatever problem we might be facing, and the more easily we recognize that whatever we experience—attachment, aversion, stress, anxiety, fear, or longing—is simply a fabrication of our own minds.

People who have invested a sincere effort in exploring their inner wealth naturally tend to develop a certain kind of fame, respect, and credibility, regardless of their external circumstances. Their conduct in all kinds of situations inspires in others a profound sense of respect, admiration, and trust. Their success in the world has nothing to do with personal ambition or a craving for attention. It doesn't come from owning a nice car or a beautiful home, or having an important job title. It stems, rather, from a spacious and relaxed state of well-being, which allows them to see people and situations more clearly, but also to maintain a basic sense of happiness regardless of their personal circumstances.

In fact, we often hear of rich, famous, or otherwise influential people who are one day forced to acknowledge that their achievements haven't given them the happiness they expected. In spite of their wealth and power, they swim in an ocean of pain, which is sometimes so deep that suicide seems the only escape. Such intense pain results from believing that objects or situations can create lasting happiness.

If you truly want to discover a lasting sense of peace and contentment, you need to learn to rest your mind. Only by resting the mind can its innate qualities be revealed. The simplest way to clear water obscured by mud and other sediments is to allow the water to grow still. In the same way, if you allow the mind to come to rest, ignorance, attachment, aversion, and all other mental afflictions will gradually settle, and the compassion, clarity, and infinite expanse of your mind's real nature will be revealed.

PART TWO

THE PATH

A disciplined mind invites true joy.

—*The Dhammapada,*
translated by Eknath Easwaran

9

FINDING YOUR BALANCE

Rest without fixation.

—GÖTSANGPA, *Radiant Jewel Lamp,*
translated by Elizabeth M. Callahan

NOW WE WILL leave the realm of science and theory behind for a while and begin discussing practical application, which in Buddhist terms is referred to as the Path. I'd like to begin with a story I heard long ago about a man who'd been an expert swimmer in his youth and began looking for a challenge in his old age that would be as engaging as swimming had been in his early life. He decided to become a monk, thinking that just as he'd mastered the waves of the ocean, he would master the waves of his mind. He found a teacher he respected, took his vows, and began to practice the lessons his teacher gave him. As is often the case, meditation didn't come easily to him, so he went to seek advice from his teacher.

His teacher asked him to sit and meditate, so that he could observe his practice. After watching for a while, the teacher saw that the old swimmer was trying too hard. So he told the man to relax. But the swimmer found even that simple instruction hard to follow. When he tried to relax, his mind drifted and his body slumped. When he tried to focus, his mind and body became too tight. Finally the teacher asked him, "You do know how to swim, don't you?"

"Of course," the man replied. "Better than anyone."

"Does the ability to swim come from holding your muscles completely tense," his teacher asked, "or completely loose?"

"Neither," the old swimmer answered. "You have to find a balance between tension and relaxation."

"Good," his teacher continued. "So now let me ask you, when you're swimming, if your muscles are too tense, are *you* creating the tightness in your limbs, or is someone else forcing you to tense up?"

The man thought a bit before answering. Finally he said, "No one outside of me is forcing me to tighten my muscles."

The teacher waited a moment for the old swimmer to absorb his own answer. Then he explained, "If you find your mind becoming too tight in meditation, you yourself are creating the tension. But if you let go of all tension, your mind becomes too loose and you become drowsy. As a swimmer, you learned how to find the proper muscular balance between tension and looseness.

"In meditation, you need to find the same equilibrium in your mind. If you don't find that equilibrium, you'll never be able to realize the perfect balance within your own nature. Once you discover the perfect balance within your own nature, you'll be able to swim through every aspect of your life the way you swim through water."

In very simple terms, the most effective approach to meditation is to try your best without focusing too much on the results.

WISDOM AND METHOD

> *When the mind is not altered, it is clear.*
> *When water is not disturbed, it is transparent.*
>
> —The Ninth Gyalwang Karmapa,
> *Mahāmudrā: The Ocean of Definitive Meaning,*
> translated by Elizabeth M. Callahan

The specific instructions the teacher gave the old swimmer were actually part of a larger lesson in finding a balance between wisdom, or philosophical understanding, and method, the practical application of philosophy. Wisdom is pointless without a practical means of applying it. That's where method comes in: using the mind to recognize the mind. That's actually a good working definition of meditation. Meditation is not about "blissing out," "spacing out," or "getting clear"—among the many terms I've heard from people in my travels around the world. Meditation is actually a very simple exercise in resting in the natural state of your present mind, and allowing yourself to be simply and clearly present to whatever thoughts, sensations, or emotions occur.

Many people resist the idea of meditation because the image that first comes to mind involves hours and hours of sitting ramrod straight, with legs crossed, and an absolutely blank mind. None of this is necessary.

First of all, sitting with your legs crossed and your spine straight takes some getting used to—especially in the West, where it's common to slouch in front of a computer or a TV. Second, it's impossible to keep your mind from generating thoughts, feelings, and sensations. Thinking is the mind's natural function, just as it's the natural function of the sun to produce light and warmth, or of a thunderstorm to produce lightning and rain. When I first began learning about meditation, I was taught that trying to suppress the natural function of my mind was at best a temporary solution. At worst, if I deliberately tried to *change* my mind, I'd really just be reinforcing my own tendency to fixate on thoughts and feelings as inherently real.

The mind is always active, always generating thoughts, just as the ocean constantly generates waves. We can't stop our thoughts any more than we can stop waves in the ocean. Resting the mind in its natural state is very different from trying to stop thoughts altogether. Buddhist meditation does not in any way involve attempting to make the mind a blank. There's no way to achieve thoughtless meditation. Even if you could manage to stop your thoughts, you wouldn't be meditating; you'd just be drifting in a zombielike state.

On the other hand, you may find that as soon as you look at a thought, an emotion, or a sensation, it vanishes like a fish suddenly swimming away to deeper waters. That's okay, too. In fact, it's great. As long as you're maintaining that sense of bare attention or awareness, even when thoughts, feelings, and so on elude you, you're experiencing the natural clarity and emptiness of your mind's true nature. The real point of meditation is to rest in bare awareness whether anything occurs or not. Whatever comes up for you, just be open and present to it, and let it go. And if nothing occurs, or if thoughts and so on vanish before you can notice them, just rest in that natural clarity.

How much simpler could the process of meditation be?

Another point to consider is that although we cling to ideas that some experiences are better, more appropriate, or more productive than others, there are, in fact, no *good* thoughts or *bad* thoughts. There are only thoughts. As soon as one bunch of gossipy neurons starts producing signals that we translate as thoughts or feelings, another group starts commenting, "Oh, *that* was a vengeful thought. What a *bad* person you are" or "You're so *afraid,* you must *really* be incompetent." Meditation is really a process of nonjudgmental awareness. When we meditate, we adopt the objective perspective of a scientist toward our own subjective experience. This might not be easy at first. Most of us are trained to believe that if we think something is good, it *is* good, and if we think something is bad, it *is* bad. But as we practice simply watching our thoughts come and go, such rigid distinctions begin to break down. Common sense will tell us so many mental events arising and vanishing in the space of a minute can't *all* possibly be true.

If we continue to simply allow ourselves to be aware of the activity of our minds, we'll very gradually come to recognize the transparent nature of the thoughts, emotions, sensations, and perceptions we once considered solid and real. It's as though layers of dust and dirt were slowly being wiped away from the surface of a mirror. As we grow accustomed to looking at the clear surface of our minds, we can see through all the gossip about who and what we think we are, and recognize the shining essence of our true nature.

PHYSICAL POSTURE

> *Great wisdom abides in the body.*
>
> —*The Hevajra Tantra,*
> translated by Elizabeth M. Callahan

The Buddha taught that the body is the physical support for the mind. The relationship between them is like the relationship between a glass and the water it contains. If you set a glass down on the edge of a table or on top of something that isn't flat, the water will shift around or possibly spill. But if you set the glass on a flat, stable surface, the water in it will remain perfectly still.

Similarly, the best way to allow the mind to come to rest is to create a stable physical posture. In his wisdom, the Buddha provided instructions for aligning the body in a balanced way that allows the mind to remain *relaxed* and *alert* at the same time. Over the years, this physical alignment has become known as the seven-point posture of Vairochana, an aspect of the Buddha that represents enlightened form.

The first point of the posture is to create a stable basis for the body, which means, if possible, crossing your legs so that each foot rests on the opposite thigh. If you can't do this, you can just cross one foot on top of the opposite thigh, resting the other beneath the opposite thigh. If neither position is comfortable, you can simply cross your legs. You can even sit comfortably in a chair, with your feet resting evenly on the floor. The goal is to create a physical foundation that is simultaneously comfortable and stable. If you feel great pain in your legs, you won't be able to rest your mind because you'll be too preoccupied by the pain. That's why there are so many options available concerning this first point.

The second point is to rest your hands in your lap just below your navel, with the back of one hand resting in the palm of the other. It doesn't matter which hand is placed on top of the other, and you can switch their positions at any time during your practice—if, for instance, the covered palm gets hot after a long time. It's also fine to simply lay your hands palm-down over your knees.

The third point is to allow a bit of a space between your upper arms and your torso. The classic Buddhist texts call this "holding your arms like a vulture," which can easily be mistaken for stretching out your shoulder blades as if you were some sort of predatory bird.

In fact, one day, while I was teaching in Paris, I happened to be walking through a park when I saw a man sitting cross-legged on the ground, repeatedly flapping his shoulders forward and back. As I passed him, he recognized that I was a monk (the red robes are pretty much a giveaway), and asked me, "Do you meditate?"

"Yes," I replied.

"Do you have any problems?" he asked.

"Not really," I told him.

We stood for a moment smiling at each other—it was a nice, sunny day in Paris, after all—and then he said, "I like meditation very much, but there's one instruction that really drives me crazy."

Naturally, I asked him what it was.

"It's the position of the arms," he replied, a little embarrassed.

"Really?" I replied. "Where did you learn about meditation?"

"From a book," he answered.

I asked him what the book had said about the position of the arms.

"It said you should hold your arms like vulture's wings," he replied—at which point he started flapping his shoulders back and forth, as I'd seen him do when I first approached. After watching him flap for a couple of seconds, I asked him to stop.

"Let me tell you something," I said. "The real point of that instruction is to keep a little bit of space between your arms and your torso, just enough to make sure that your chest is open and relaxed, so that you can breathe nice and freely. Vultures at rest always have a little bit of space between their wings and their bodies. That's really what the instruction means. There's no need to flap your arms. After all, you're just trying to meditate. You're not trying to fly."

The essence of this point of physical posture is to find a balance between your shoulders so that one is not dipping below the other, while keeping your chest open to allow yourself some "breathing room." Some people have very big arms or very big torsos—especially if they've spent a lot of time working out at the gym. If you happen to fall into this category, don't strain yourself to artificially maintain a bit of space between your arms and chest. Just allow your arms to rest naturally in a way that doesn't constrict your chest.

The fourth point of the physical posture is to keep your spine as straight as possible—as the classic texts say, "like an arrow." But here, again, it's important to find a balance. If you try to sit up *too* straight, you'll end up leaning backward, your whole body shaking with tension. I've seen this happen many times with students who are overly concerned with having an absolutely erect spine. On the other hand, if you just let yourself slouch, you'll almost certainly end up compressing your lungs, which will make it harder to breathe, as well as squashing various internal organs, which can be a source of physical discomfort.

The fifth point involves letting the weight of your head rest evenly on your neck, so that you're not crushing your windpipe or straining so far backward that you compress the cervical vertebrae, the seven little bones at the top of your spinal cord, which is vital in transmitting neuronal signals from the lower parts of your body to your brain. When you find the position that's right for you, you'll probably notice that your chin is tilting just slightly more toward your throat than it ordinarily does. If you've ever sat in front of a computer for hours and hours with your head tipped slightly backward, you'll immediately understand how much better you'll feel by making this simple adjustment.

The sixth point concerns the mouth, which should be allowed to rest naturally so that your teeth and lips are very slightly parted. If possible, you can allow the tip of your tongue to gently touch the upper palate just behind the teeth. Don't force the tongue to touch your palate; just allow it to rest there gently. If your tongue is too short to reach the palate without strain, don't worry. The most important thing is to allow the tongue to rest naturally.

The last point of the meditation posture involves the eyes. Most people who are new to meditation feel more comfortable keeping their eyes closed. They find it easier that way to allow the mind to rest and to experience a sense of peace and tranquillity. This is fine at the beginning. One of the things I learned early on, however, is that keeping the eyes closed makes it easier to become attached to an artificial sense of tranquillity. So, eventually, after a few days of practice, it's better to keep your eyes open when you meditate, so that you can stay alert, clear, and mindful. This doesn't mean glaring straight ahead without blinking, but simply leaving your eyes open as they normally are throughout the day.

The seven-point posture of Vairochana is really a set of guidelines. Meditation is a personal practice, and everyone is different. The most important thing is to find *for yourself* the appropriate balance between tension and relaxation.

There's also a short, two-point meditation posture, which can be adopted at times when it may be inconvenient or impossible to settle fully into the more formal seven-point posture. The instructions are very simple: Just keep your spine straight and the rest of your body as loose and relaxed as possible. The two-point meditation posture is very useful throughout the day, while going about your daily activities like driving, walking down the street, grocery shopping, or making dinner.

This two-point posture in itself almost automatically produces a sense of relaxed awareness—and the best part is that when you assume it, no one will even notice that you're meditating at all!

MENTAL POSTURE

> *If the mind itself that is twisted into knots is loosened, it is undoubtedly freed.*
>
> —Saraha, Doha for the People,
translated by Elizabeth M. Callahan

The same principles behind finding a relaxed and alert physical posture apply to finding the same sort of balance within your mind. When your mind is poised naturally between relaxation and alertness, its innate qualities spontaneously emerge. This was one of the things I learned during those three days I spent sitting alone in my retreat room, determined to observe my mind. As I sat there, I kept remembering how my teachers had told me that when water becomes still, the silt, mud, and other sediment gradually separates from the water and settles to the bottom, giving you a chance to see the water and whatever passes through it very clearly. In the same way, if you remain in a state of mental relaxation, the "mental sediment" of thoughts, emotions, sensations, and perceptions naturally subsides and the mind's innate clarity is revealed.

Just as in the case of physical posture, the essential point of mental posture is to find a balance. If your mind is too tight or too focused, you'll end up becoming anxious over whether you're being a *good* meditator. If your mind is too loose, you'll either get carried away by distractions or fall into a kind of dullness. You want to find a middle way between perfection-driven tightness and a kind of disenchanted "Oh no, I've got to sit down and meditate" type of dreariness. The ideal approach is to give yourself the freedom to remember that whether your practice is good or not doesn't really matter. The important point is the intention to meditate. That alone is enough.

10

SIMPLY RESTING: THE FIRST STEP

Look naturally at the essence of whatever occurs.
—KARMA CHAGMEY RINPOCHE,
The Union of Mahamudra and Dzogchen,
translated by Erik Pema Kunsang

THE BUDDHA RECOGNIZED that no two people are exactly alike, that everyone is born with a unique combination of abilities, qualities, and temperaments. It is a measure of his great insight and compassion that he was able to develop an enormous variety of methods through which all sorts of people might arrive at a direct experience of their true nature and become completely free from suffering.

Most of what the Buddha taught was delivered spontaneously according to the needs of the people who happened to be around him at any given moment. The ability to respond spontaneously in precisely the right way is one of the marks of an enlightened master—which works quite nicely as long as the enlightened master happens to be alive. After the Buddha died, however, his earliest students had to figure out a way to organize these spontaneous teachings in a way that would be useful to the generations that would follow. Fortunately, the Buddha's early followers were very good at creating classifications and categories, and came up with a way to organize the various meditation practices the Buddha taught into two basic categories: analytical methods and nonanalytical methods.

The nonanalytical methods are usually taught first, because they provide the means for calming the mind. When the mind is calm, it's much easier to simply be aware of various thoughts, feelings, and sensations without getting caught up in them. The analytical practices involve looking directly at the mind in the midst of experience, and are usually taught after someone has had some practice in learning how to rest the mind simply as it is. Also, because the experience of looking directly at the mind can provoke a lot of questions, the analytical practices are best undertaken under the supervision of a teacher who has the insight and experience to understand these questions and provide answers that are uniquely suited to each student. For this reason, the meditation practices I want to focus on here are the ones related to resting and calming the mind.

In Sanskrit, the nonanalytical approach is known as *shamata*. In Tibetan, it is called *shinay,* a word made up of two syllables: *shi,* which means "peace" or "tranquillity," and *nay,* which means "to abide" or "stay." Translated into English, then, this approach is known as *calm abiding*—simply allowing the mind to rest calmly as it is. It's a basic kind of practice through which we rest the mind naturally in a state of relaxed awareness in order to allow the nature of mind to reveal itself.

OBJECTLESS MEDITATION

> *Cut through the root of your own mind: Rest within*
> *naked awareness.*
>
> —TILOPA, *Ganges Mahāmudrā,*
> translated by Elizabeth M. Callahan

When my father first taught me about resting the mind naturally, "within naked awareness," I had no idea what he was talking about. How was I supposed to just "rest" my mind without something to rest it *on?*

Fortunately, my father had already traveled a bit around the world and had met quite a few people and was able to strike up some conversations with them about their lives, their problems, and their successes. This is actually one of the great advantages of wearing Buddhist robes. People are more inclined to think you're wise or important, and willingly just open up and start telling you details about their lives.

The example my father used about resting the mind came from something he'd heard from a hotel clerk, who was always happy to end his day, which entailed standing behind a desk for eight hours, checking people in, checking them out, listening to their complaints about their rooms, and endlessly arguing about charges on their bills. At the end of his shift, the clerk was so physically exhausted that all he looked forward to was going home and sitting in a nice, long bath. And after his bath, he'd go to his bedroom, rest on his bed, let out a sigh, and just relax. The next few hours were his alone: no standing on his feet in a uniform, no listening to complaints, and no staring at the computer to confirm reservations and look up room availabilities.

That's how to rest the mind in objectless shinay meditation: as though you've just finished a long day of work. Just let go and relax. You don't have to block whatever thoughts, emotions, or sensations arise, but neither do you have to follow them. Just rest in the open present, simply allowing whatever happens to occur. If thoughts or emotions come up, just allow yourself to be aware of them. Objectless shinay meditation doesn't mean just letting your mind wander aimlessly among fantasies, memories, or daydreams. There's still some presence of mind that may be loosely described as a center of awareness. You may not be fixating on anything in particular, but you're still aware, still present to what's happening in the here and now.

When we meditate in this objectless state, we're actually resting the mind in its natural clarity, entirely indifferent to the passage of thoughts and emotions. This natural clarity—which is beyond any dualistic grasping of subject and object—is always present for us in the same way that space is always present. In a sense, objectless meditation is like accepting whatever clouds and mist might obscure the sky while recognizing that the sky itself remains unchanged even when it is obscured. If you've ever flown in an airplane, you've probably witnessed that above any clouds, mist, or rain, the sky is always open and clear. It looks so ordinary. In the same way, Buddha nature is always open and clear even when thoughts and emotions obscure it. Though it may seem very ordinary, all the qualities of clarity, emptiness, and compassion are contained within that state.

Objectless shinay practice is the most basic approach to resting the mind. You don't have to watch your thoughts or emotions—practices that I will discuss later on—nor do you have to try to block them. All you need to do is rest within the awareness of your mind going about its business with a kind of childlike innocence, a sense of "Wow! Look how many thoughts, sensations, and emotions are passing through my awareness right now!"

In a sense, objectless shinay practice is similar to looking at the vast expanse of space rather than focusing on the galaxies, stars, and planets that move through it. Thoughts, emotions, and sensations come and go in awareness, the way galaxies, stars, and planets move through space. Just as space isn't defined by the objects that move through it, awareness isn't defined or limited by the thoughts, emotions, perceptions, and so on that it apprehends. Awareness simply *is*. And objectless shinay practice involves simply resting in the "is-ness" of awareness. Some people find the practice quite easy; others find it very difficult. It's more a matter of individual temperament than of competence or skill.

The instructions are simple. If you're practicing formally, it's best to assume the seven-point posture to the best of your ability. If you can't assume a formal posture—if you're driving, for example, or walking down the street—then simply straighten your spine while keeping the rest of your body relaxed and balanced. Then allow your mind to relax in a state of bare awareness of the present.

Inevitably, all sorts of thoughts, sensations, and feelings will pass through your mind. This is to be expected, since you haven't trained in resting the mind. It's just like starting a weight-training program at the gym. At first you can lift only a few pounds for a few repetitions before your muscles get tired. But if you keep at it, gradually you'll find that you can lift heavier weights and perform more repetitions.

Similarly, learning to meditate is a gradual process. At first you might be able to remain still for only a few seconds at a time before thoughts, emotions, and sensations bubble up to the surface. The basic instruction is simply not to follow after these thoughts and emotions, but merely to be aware of everything that passes through your awareness as it is. Whatever passes through your mind, don't focus on it and don't try to suppress it. Just observe it as it comes and goes.

Once you begin following after a thought, you lose touch with what's happening in the here and now, and you begin imagining all sorts of fantasies, judgments, memories, and other scenarios that may have nothing to do with the reality of the present moment. And the more you allow yourself to get caught up in this type of mental wandering, the easier it becomes to drift away from the openness of the present moment.

The purpose of shinay meditation is to slowly and gradually break this habit and remain in a state of present awareness—open to all the possibilities of the present moment. Don't criticize or condemn yourself when you find yourself following after thoughts. The fact that you've caught yourself reliving a past event or projecting into the future is enough to bring you back to the present moment and strengthens your intention to meditate. Your *intention to meditate* as you engage in practice is the crucial factor.

It's also important to proceed slowly. My father was very careful to tell all his new students, including me, that the most effective approach in the beginning is to rest the mind for very short periods many times a day. Otherwise, he said, you run the risk of growing bored or becoming disappointed with your progress and eventually give up trying altogether. "Drip by drip," the old texts say, "a cup gets filled."

So, when you first start out, don't set yourself a lofty goal of sitting down to meditate for twenty minutes. Aim instead for a minute or even half a minute—utilizing those few seconds when you find yourself willing or even desiring just to take a break from the daily grind to observe your mind rather than drifting off into daydreams. Practicing like this, "one drip at a time," you'll find yourself gradually becoming free of the mental and emotional limitations that are the source of fatigue, disappointment, anger, and despair, and discover within yourself an unlimited source of clarity, wisdom, diligence, peace, and compassion.

11

NEXT STEPS:
RESTING ON OBJECTS

Rest the mind by directing one-pointed attention on a specific object.

—The Ninth Gyalwang Karmapa,
    Mahāmudrā: The Ocean of Definitive Meaning,
    translated by Elizabeth M. Callahan

WHEN I FIRST started meditating formally, I found that objectless meditation was too hard because it was too easy. The simple awareness that is the essence of natural mind is too close to recognize. It's there when we wake up in the morning, wherever we go throughout the day, when we eat, and when we get ready to go to bed. It's simply awareness. It is what it is. But because it's with us all the time, we don't recognize how precious it is. And it's just too easy to get caught up in all the thoughts, feelings, and sensations that are the natural by-products of the mind in its natural state.

If you find yourself facing this problem, you're not alone.

Fortunately, my father and my other teachers were well acquainted with the problem of resting the mind directly and were able to teach other, more gradual techniques. The simplest methods involved directly using the senses as a means of calming and relaxing the mind.

THE DOORS OF PERCEPTION

This whole world is mind's world, the product of mind.
　　　　　　　　　—CHÖGYAM TRUNGPA, *The Heart of the Buddha*

Like scientists, Buddhists recognize the five senses of sight, hearing, smell, taste, and touch. In Buddhist terms, the five sense faculties are known as the doors of perception, an image based on the openings of a house. Most of our feelings and perceptions enter our experience through one or more of these five doors. But since the five sense faculties—or sense consciousnesses, as most Buddhist texts refer to them—can only register sensory perceptions, Buddhist science adds a sixth sense, the mental consciousness. There's nothing mysterious or occult about this sixth consciousness. It has nothing to do with extrasensory perception or being able to talk to spirits. It's simply the capacity of mind to know and discern what we see, smell, hear, taste, or touch.

The traditional metaphor for the six consciousnesses is a house with five openings, one in each of the four directions and one in the roof. These five openings represent the five sense consciousnesses. Now suppose someone sets a monkey loose in this building. The monkey represents the mental consciousness. Suddenly set free in a big house, the monkey would naturally go crazy jumping around from opening to opening to check things out, looking for something new, something different, something interesting. Depending on what it finds, this crazy monkey decides whether an object it perceives is pleasurable or painful, good or bad, or in some cases simply boring. Anybody passing by the house and seeing a monkey at every opening might think there are five monkeys loose in the house. Really, though, there's only one: the restless, untrained mental consciousness.

But like every other sentient being, all a crazy monkey really wants is to be happy and to avoid pain. So it's possible to teach the crazy monkey in your own mind to calm down by deliberately focusing its attention on one or another of the senses.

OBJECT MEDITATION

To counteract our tendency to constantly fabricate, the Buddhas taught us to rely on a support. By getting accustomed to this support, our attention becomes stabilized.

—TULKU URGYEN RINPOCHE, *As It Is*, Volume 1, translated by Erik Pema Kunsang

In the course of ordinary experience, the information we receive from our senses is almost inevitably a source of distraction, since the mind tends to fixate on sensory information. At the same time, because we're embodied beings, we would inevitably experience a sense of futility if we attempted to disengage completely from our senses or block the information we receive through them. The more practical approach is to make friends with our senses, and to utilize the information we receive through our sensory organs as a means of calming the mind. Buddhist texts refer to this process as "self-antidote," using the source of distraction itself as the means to attaining freedom from distraction. The metaphor derives from the practice common in ancient times of using the same material to work with a particular substance. If you wanted to cut glass, for example, you had to use glass; if you wanted to cut through iron, you had to use a tool made of iron. In the same way, you can use your senses to cut through the distractions of the senses.

In object meditation practice, we use our senses as a means of stabilizing the mind. We can use the faculty of sight to meditate on form and color; the faculty of hearing to meditate on sound; the faculty of smell to meditate on odors; the faculty of taste to meditate on flavors; and the faculty of touch to meditate on physical sensations. Instead of distractions, the information we receive through our senses can become great assets to our practice.

Once I learned how to observe my perceptions in a calm, meditative way, practice became a whole lot easier. I found myself getting far less emotionally involved in what I perceived. Instead of thinking, *Oh, this guy is yelling at me,* I was able to think, *Hmm, this guy's voice is pretty loud, his tone's a bit sharp, and the sounds he's making are probably intended to be insulting or hurtful.*

In other words, just through learning how to rest my attention very lightly on the sensory information I was receiving, and disengaging from the emotional or intellectual content normally associated with the sounds he was making, *he couldn't hurt me.* And by being able to listen to him nondefensively, I found myself open enough to respond in a way that defused his apparent anger without diminishing my own integrity.

Meditating on Physical Sensations

One of the easiest ways to begin object-based shinay practice is to rest your attention gently on simple, physical sensations. Just focus your attention on a particular area—for example, your forehead.

Start by straightening your spine and relaxing your body. If you're practicing formally, you can assume the seven-point posture described earlier. If you're someplace where it's not convenient to assume the posture, just straighten your spine and allow the rest of your body to relax comfortably. It doesn't matter whether you keep your eyes open or closed as you practice. In fact, some people find it more helpful to keep their eyes closed. (Of course, if you're driving or walking down the street, I would strongly suggest you keep your eyes open!)

Let your mind rest for a few moments, just as it is. . . .

Now slowly bring your awareness to your forehead. . . .

You might feel a sort of tingling there, or maybe a sensation of warmth. You might even feel some sort of itchiness or pressure. Whatever you feel, just allow yourself to be aware of it for a minute or two. . . .

Just notice it. . . .

Just gently rest your attention on the sensation. . . .

Then let go of your attention and let your mind rest as it is. If you've had your eyes closed, you can open them.

How was that?

After you've spent a little time resting your awareness on the sensations in one part of the body, you can extend the technique by gently drawing your attention throughout your entire body. I sometimes refer to this extended approach to physical sensations as "scanning practice," because it reminds me of lying in one of those machines that can scan your entire body. Again, if you're practicing formally, start by adopting the seven-point posture. If you're practicing informally, just straighten your spine and let the rest of your body relax comfortably and naturally. In either case, you can keep your eyes open or closed, whichever is most comfortable for you.

Begin by just allowing your mind to rest in objectless shinay for a few moments. Then gently bring your awareness to whatever sensations you feel in your forehead region. Allow your mind to simply observe these sensations, to be simply aware, nothing more. Gradually lower your focus, observing whatever sensations are occurring in your face, your neck, your shoulders, your arms, and so on. Just observe. There's no need to block anything else from your mind, or to change what you observe. Just keep your mind and body relaxed and quiet while simply recognizing sensations as they arise. After a few minutes, allow your mind simply to rest. Then return to observing your sensations, alternating between observing and resting the mind for as long as your practice session lasts.

Most sensations involve some sort of physical basis. Our bodies come into contact with something: the chair we're sitting on, the floor, a pen, our clothes, an animal, or a person. And that contact produces a distinct physical sensation. In Buddhist terms, the kinds of sensations that result from direct physical contact are referred to as "gross physical sensations." But as we become more deeply attentive to what we feel, we begin to recognize feelings that aren't necessarily related to tactile contact, feelings that are referred to as "subtle physical sensations."

When I first began to practice this sort of shinay technique, I discovered that when I tried to avoid a particular sensation, it increased. But when I learned to just look at it, whatever discomfort I felt became more tolerable. Being a curious child, of course, I had to know why this shift occurred. Only after watching the process for a while did I realize that when I simply allowed myself to observe a sensation, I was actively participating in what was occurring right then, at that moment. I saw part of my mind resisting a painful sensation and part of my mind urging me to just look at it objectively. When I looked at these conflicting impulses simultaneously, I was able to see my whole mind engaged in the process of dealing with avoidance and acceptance, and the process of observing the workings of my mind became more interesting than either avoidance or acceptance. Just watching my mind work was fascinating in itself. This, I think, is the most practical definition of clarity I can offer: the capacity to see the mind working simultaneously on many levels.

Meditating on Painful Sensations

Feelings like being cold, hot, hungry, full, heavy, or dizzy, or having a headache, a toothache, a stuffy nose, a sore throat, or pain in your knees or lower back, are pretty much directly—though not always pleasantly—present to awareness. Because pain and discomfort are such direct sensations, they're actually very effective objects of meditative focus. Most of us regard pain as a threat to our physical well-being. On one hand, when we worry or allow ourselves to become preoccupied by this threat, the pain itself almost always increases. On the other hand, if we consider pain or discomfort as an object of meditation, we can use such sensations to increase our capacity for clarity, simply through watching the mind deal with various solutions.

For example, if I feel some pain in my legs or lower back while sitting in formal meditation or even while just sitting in a car or an airplane, instead of stretching, getting up, or moving around, I've learned to look at the mental experience of pain. It's the mental consciousness, after all, that actually recognizes and registers sensations. When I bring my attention to the *mind* that is registering pain, rather than focusing on a particular area of pain, the pain doesn't necessarily disappear, but it becomes a point of actively engaging with whatever I'm experiencing here and now, rather than trying to avoid it. The same principle holds true for pleasurable sensations: Rather than trying to sustain them, I just observe them as manifestations of experience. In effect, my early years of training began to show me how to use sensations as a means of examining and appreciating the infinite capacity of the mind, rather than being used by sensations to enforce a sense of being bound by physical limitations.

Of course, if you're experiencing chronic or serious pain you should consult a doctor, as these symptoms may indicate a serious physical problem. I've heard from some people, however, that after their doctors have ruled out any serious medical problem, the pain they experienced actually subsides. It seems as if fear of pain exacerbates the sensation of pain and "locks" it into place—which may represent a self-perpetuating "red alert" signal sent from the thalamus to the amygdala and other parts of the brain. If your doctor has uncovered a serious medical problem, however, by all means follow his or her recommendations for treatment. Although meditation can help you to deal with the pain and discomfort of serious medical problems, it is not a substitute for treatment.

Even while taking prescribed or over-the-counter medications recommended by a doctor, you may experience some pain, in which case you can try working with the physical sensation of pain as a support for meditation. If the pain you experience is a symptom of a serious medical condition, avoid focusing on results. If your underlying motivation is to get rid of the pain, you're actually reinforcing the neuronal patterns associated with the fear of pain. The best way to weaken these neuronal patterns is to simply make the effort to observe the pain objectively, leaving results to sort themselves out.

I was never so impressed by this lesson as when my father had to undergo a minor operation while he was in Germany. Apparently, whoever was supposed to anesthetize the area to be operated on got caught up in other duties and completely forgot about my father. When the doctor made the first incision, he noticed that the muscles in the area began to twitch—which wouldn't have happened if the area had been properly numbed. The doctor was furious with the anesthetist, but my father begged him not to cause any trouble, because he didn't feel any pain. The sensation of having such a sensitive area incised, he explained, had actually provided him with an opportunity to raise his awareness to a heightened sense of clarity and peace.

In simple terms, my father had developed through practice a network of neuronal connections that spontaneously fired to elevate the experience of pain to an objective observation of the mind that experiences pain. Though the doctor did insist on anesthetizing the area before continuing the operation, at my father's insistence he ended up not lodging a complaint against the woman who was supposed to have administered the anesthesia.

The next day the anesthetist came to my father's bedside, holding something behind her back. Smiling, she thanked him for keeping her out of trouble, and then produced from behind her back a bag full of treats, which he found quite delicious.

The practice of observing physical sensations—whether "gross" or "subtle"—is so simple that you can use it either during formal meditation sessions or at any point during the day when you find yourself with a few spare seconds between meetings, appointments, or other obligations. In fact, I've found this practice especially useful throughout the day because it generates an immediate feeling of lightness and openness. Several people have told me they've found the practice quite useful at work, when they find themselves sitting for hours listening to a boring presentation.

Meditating on Form

The technical name for using the sense of sight as a means for resting the mind is "form meditation." But don't let the name scare you. Form meditation is actually very simple. In fact, we practice it unconsciously every day whenever we stare at a computer screen or watch a traffic light. When we lift this unconscious process to the level of active awareness, deliberately resting our attention upon a specific object, the mind becomes very peaceful, very open, and very relaxed.

I was taught to start with a very small object located close enough to see without strain. It could be a patch of color on the floor, a candle flame, a photograph, or even the back of the head of the person sitting in front of me in a classroom. It's also fine to look at an object of more spiritual significance—often referred to as a "pure form." If you're a Buddhist, the object might be an image or statue of the Buddha; if you're a Christian, you might focus on a cross or an image of a saint; if you belong to some other religious tradition, choose an object that has special significance for you. As you become more familiar with this practice, it's even possible to focus on mental forms—objects recalled simply in your imagination.

Whatever object you choose, you'll probably notice that it has two characteristics: shape and color. Focus on whichever aspect you prefer. You could choose something white, black, or pink, or round, square, or multiform. The object itself doesn't matter. The idea is simply to rest your attention on either its color or its shape, engaging the mental faculty only to the point of barely recognizing the object. Nothing more than that. The moment you bring attention to the object, you are aware.

It's not necessary to try to see it so clearly that you recognize every little detail. If you try to do that, you'll tense up, whereas the whole point of the exercise is to rest. Keep your focus loose, with just enough attention to hold the bare awareness of the object you're looking at. Don't try to *make* anything happen or try to force your mind to relax. Simply think, *Okay, whatever happens, happens. This is meditation. This is what I'm doing.* It doesn't have to be anything more than that.

Of course, it's possible to stare open-eyed at an object without really seeing it at all. Your mind might be completely absorbed by something you hear in the distance, so for several seconds or even minutes you don't see the object at all. How I hated it when my mind drifted like that! But, according to my father, this sort of drifting is entirely natural. And when you recognize that your mind has drifted away from the object of focus, just bring your attention back to the object.

So now I encourage you to practice.

Assume whatever physical posture is most comfortable for you and just allow your mind to rest for a few moments in a very relaxed, loose state. Then choose something to look at and just rest your gaze on it, noticing its shape or its color. You don't have to stare intently—if you need to blink, just blink. In fact, if you don't blink, your eyes will likely become very dry and irritated. After a few moments of gazing at the object, let your mind simply relax again. Return your focus to the object for a few minutes; then allow your mind to relax once more.

Whenever I practice using a visual object as a support, I'm reminded of something mentioned by Longchenpa, one of the great Buddhist scholars and meditation masters of the fourteenth century. In one of his books he points out that there is a great benefit to be gained from alternating between object-based meditation and the sort of objectless meditation discussed earlier. As he explains, when you rest your mind on an object, you're seeing it as something distinct or separate from yourself. But when you let go and simply rest your mind in naked awareness, the distinction dissolves. And alternating between focusing on an object and allowing the mind to rest in naked awareness, you actually come to recognize the basic truth that neuroscience has shown us: Everything we perceive is a reconstruction created in the mind. In other words, there's no difference between what is seen and the mind that sees it.

This recognition doesn't, of course, occur overnight. It takes a bit of practice. In fact, as we'll see later on, the Buddha provided some specific methods for dissolving the distinction between the mind and whatever the mind perceives. But I'm getting ahead of myself—which happens when I get excited about something. For now, let's go back to basic methods of transforming sensory information into a means of bringing the mind to a calm, restful place.

Meditating on Sound

Meditating on sound is very similar to meditating on form, except that now you're engaging the faculty of hearing. Start by just allowing your mind to rest for a few moments in a relaxed state, and then gradually allow yourself to become aware of the things you hear close to your ear, such as your heartbeat or your breath, or sounds that occur naturally in your immediate surroundings. Some people find it helpful to play a recording of natural sounds or pleasant music. There's no need to try to identify these sounds, nor is it necessary to focus on a specific sound. In fact, it's easier to let yourself be aware of everything you hear. The point is to cultivate a simple, bare awareness of sound as it strikes your ear.

As with form and color meditation, you'll probably find that you can focus on the sounds around you for only a few seconds at a time before your mind wanders off. That's okay. When you find your mind wandering, just bring yourself back to a relaxed state of mind and then bring your awareness back to the sound once again. Allow yourself to alternate between resting your attention on sounds and allowing your mind simply to rest in a relaxed state of open meditation.

One of the great benefits of meditation on sound is that it gradually teaches you to detach from assigning *meaning* to the various sounds you hear. You learn to listen to what you hear without necessarily responding emotionally to the *content*. As you grow accustomed to giving bare attention to sound simply *as* sound, you'll find yourself able to listen to criticism without becoming angry or defensive and able to listen to praise without becoming overly proud or excited. You can simply listen to what other people say with a much more relaxed and balanced attitude, without being carried away by an emotional response.

I once heard a wonderful story about a famous sitar player in India who learned to use the sounds of his instrument as a support for his meditation practice. If you're not familiar with Indian instruments, a sitar is a very long-necked instrument, usually constructed with seventeen strings, plucked like a guitar to produce a wonderful variety of tones. This particular sitar player was so gifted that he was always in demand and spent much of his time traveling around India, in much the way some modern rock bands are often away from home on tour.

After one particularly long tour, he returned home to discover that his wife had been having an affair with another man. He was extraordinarily reasonable when he discovered the situation. Perhaps the concentration he'd learned over the years of constant practice and performance, combined with the sounds of this lovely instrument, had calmed and focused his mind. In any case, he didn't argue with his wife or lash out in anger. Instead he sat down and had a long conversation with her, during which he realized that his wife's affair and his own pride at being asked to perform across the country were symptoms of attachment—one of the three mental poisons that keep us addicted to the cycle of samsara. There was very little difference between his attachment to being famous and his wife's attachment to another man. The recognition hit him like a thunderbolt, and he realized that in order to become free of his own addiction, he had to let go of his attachment to being famous. The only way for him to do so was to seek out a meditation master and learn how to recognize his attachment as simply a manifestation of his mental habits.

At the end of the conversation, he gave up everything to his wife except his sitar, toward which he still felt a strong attachment that no amount of rational analysis could dissolve, and went in search of a teacher. Eventually he arrived at a charnel ground, the ancient equivalent of a cemetery, in which corpses are more or less deposited without being buried or cremated. Charnel grounds were scary places, covered with human bones, partial skeletons, and rotting corpses. But they were the most likely environments in which to find a great master, who had overcome his or her fear of death and impermanence—two of the fearful conditions that keep most people locked in the samsaric conditions of attachment to what is and aversion to what might occur.

In this particular charnel ground, the sitar player found a *mahasiddha*—a person who had passed through extraordinary trials to achieve profound understanding. The mahasiddha was living in a ragged hut that barely provided him protection against wind and weather. In the way that some of us feel a strong connection with people we meet during the ordinary course of our lives, the sitar player felt a deep bond with this particular mahasiddha and asked him if he would accept him as a student. The mahasiddha agreed, and the sitar player used branches and mud to build his own hut nearby, where he could practice the basic instructions on shinay meditation that the mahasiddha had given him.

Like many people who begin meditation practice, the sitar player found it very difficult to follow the instructions of his teacher. Even spending a few minutes following his teacher's instructions seemed like an eternity; every time he sat to meditate, he found himself drawn to his old habit of playing his sitar, and he gave up his practice and started to play. He began to feel horribly guilty, neglecting his meditation practice in favor of simply strumming his sitar. Finally he went to his teacher's hut and confessed that he just couldn't meditate.

"What's the problem?" the mahasiddha asked.

The sitar player replied, "I'm just too attached to my sitar. I'd rather play it than meditate."

The mahasiddha told him, "That's not a big problem. I can give you an exercise in sitar meditation."

The sitar player, who'd been expecting criticism—as most of us do from our teachers—was quite surprised.

The mahasiddha continued, "Go back to your hut, play your sitar, and just listen to the sound of your instrument with bare awareness. Forget about trying to play perfectly. Just listen to the sounds."

Relieved, the sitar player returned to his hut and started playing, just listening to the sounds without trying to be perfect, without focusing on either the results of his playing or the results of his practice. Because he'd learned to practice simply without concern for the results, after a few years he became a mahasiddha himself.

Since not many of my students are sitar players, the real lesson in this story lies in learning how to use their own experience as a support for practice, without regard to results. Especially in the West, where the sounds, sights, and smells of rush-hour traffic can become an overwhelming source of preoccupation, the practice of simply observing the sensations of traffic rather than focusing on the goal of getting through congestion offers a tremendous opportunity for meditation practice. If you turn your attention away from the goal of getting somewhere and instead rest your attention on the sensations around you, you could very well end up becoming a "traffic mahasiddha."

Meditating on Smell

Actually, we can use as an object of meditation whatever sensation draws us most strongly at any given moment. For example, using smell as an object of meditation can be especially helpful, either during formal practice or just while going about your day. In formal practice you can focus your attention on the odors around you—maybe the smell of incense, if you like that, or the smells naturally occurring around your practice area.

Meditating on smell can be especially practical if you're involved in daily activities like cooking or eating. By taking the time to focus attention on the smells that arise from food, you can transform boring daily routines—such as cooking, eating, or simply walking through your office building—into practices that calm and strengthen your mind.

Meditating on Taste

It took me quite a while to realize that when I was eating or drinking, I barely noticed what I was doing. I was usually caught up in conversations with other people, or distracted by my own problems, conflicts, or daydreams. As a result, I wasn't really engaged in what I was doing, and so I missed out on the opportunity to experience the richness of the present moment. Focusing on taste is an extremely practical technique that can be used to engage in meditation for a few moments at several points throughout the day.

When I was taught about using taste as a focus of meditation, I was told to begin as usual by allowing my mind to rest naturally for a few moments, then allow myself to focus my attention lightly on the tastes I perceived. I didn't have to analyze a particular taste sensation, like bitterness, sweetness, or sourness. I just had to rest my attention lightly on all the tastes I perceived and then rest my mind naturally, alternating between bringing my attention to the sensation of taste and resting my mind naturally.

OTHER HELPFUL SUPPORTS

> *It is to guide students well that I taught different approaches.*
> —The Lankāvatārasūtra,
> translated by Maria Montenegro

In addition to working with sense objects, the Buddha also taught a few other techniques that could be easily used anytime, anywhere. One of these techniques involves using the breath as an object of meditation. If you're alive, there's a good chance you're breathing, and the ability to direct your attention to the coming and going of the breath is always available. The second support is an old friend of mine, and one I feel especially grateful for, since it kept me from going crazy as a child. This support, with which I became acquainted purely by accident while sitting in a cave, is based on the repetition of a mantra.

Breathe In, Breathe Out

I was taught a number of different ways to use the breath as an object of meditation, but I won't bore you with them all. Instead, I'll focus on two of the simplest methods, which are also the easiest to practice without drawing attention to yourself in public. All you have to do is focus your attention lightly on the simple act of inhaling and exhaling. You can place your attention on the passage of air through your nostrils or on the sensation of air filling and exiting your lungs. Using your breath in this way is very similar to focusing on physical sensation, except that you're narrowing the awareness of sensation to the simple experience of inhaling and exhaling. Because there is a natural split-second gap between inhalation and exhalation, you can also focus on the three-part process of inhalation, exhalation, and the interval between.

Focusing on the breath is particularly useful when you catch yourself feeling stressed or distracted. Internally, the simple act of drawing attention to your breath produces a state of calmness and awareness that allows you to step back from whatever problems you might be facing and respond to them more calmly and objectively. If you're stressed out, just bring your attention to your breathing. No one will notice that you're meditating. They probably won't even pay attention to the fact that you're breathing at all.

Formal meditation on the breath is a little different. One of the methods I was taught was simply to count my inhalations and exhalations as a way of focusing my attention more completely. Count the first inhalation and exhalation as "one," the next inhalation and exhalation as "two," and so on until you reach seven. Then just start the process again from "one." Eventually, you can begin to accumulate even higher numbers of inhalations and exhalations. But, as always, it's best to begin by limiting your goal to short practice periods that you can repeat many times.

My Old Friend, Mantra

Mantra meditation is a very powerful technique that not only cultivates clear awareness, but also, through the potency of syllables that have been recited by enlightened masters for thousands of years, clears away layers of mental obscuration and increases our capacity to benefit ourselves and others. This connection may be hard to accept at first; it seems too much like magic. It might be easier to think of mantric syllables as sound waves that perpetuate through space for thousands, perhaps millions, of years.

In mantra meditation, the focus of your attention is on the mental recitation of a certain set of syllables that appear to have a direct effect on calming and clearing the mind. For this exercise, we'll use a very simple set of three syllables that make up the most basic of all mantras: OM AH HUNG. OM represents the lucid, distinctive, perceptual aspect of experience; AH represents the empty, or inherently open, aspect; while HUNG represents the union of distinctive appearance and the inherently empty nature of the appearance.

You can start by reciting the mantra aloud, and then gradually slip into a more internal form of mental recitation. The important thing is to continue reciting the mantra mentally for about three minutes, and then just let your mind rest, alternating between recitation and resting for as long as you can. Whether you feel the effects immediately or not, you've set something in motion. That "something" is the freedom of your mind.

But freedom rarely arrives in the form we think it should. In fact, for most of us, freedom feels not only unfamiliar but distinctly unpleasant. That's because we're used to our chains. They might chafe, they might make us bleed, but at least they're familiar.

Familiarity is just a thought, however, or sometimes a feeling. And to help us through the difficult transition between familiarity and freedom, the Buddha provided methods for working directly with thoughts and feelings.

12

WORKING WITH THOUGHTS AND FEELINGS

Turn your back on longing! Root out attachment!
—JAMGÖN KONGTRUL *The Torch of Certainty,*
translated by Judith Hanson

LONG AGO IN India, there was a cowherd who'd spent most of his life looking after his master's cows. Finally, at the age of sixty or so, he realized, "This is a boring job. Every day it's the same thing. Bring the cows out to pasture, watch them graze, and bring them back home. What am I supposed to learn from this?" After thinking about the matter for a while, he decided he would give up his job and learn how to meditate, so he could at least free himself from the monotony of samsara.

After giving up his job, he traveled to the mountains. One day he saw a cave, in which sat a mahasiddha. Upon seeing the mahasiddha, the cowherd felt very happy, and approached him for advice on how to meditate. The master agreed and gave the cowherd basic instructions on how to meditate using thoughts as a support. After receiving the instructions, the cowherd settled in a nearby cave and set himself to practice.

Like most of us, he ran into problems right away. During all his years as a cowherd, he'd become very fond of his cows, and when he tried to practice what the mahasiddha had taught him, the only thoughts and images that appeared in his mind were of the cows he'd taken care of. Though he tried hard to block the thoughts, the cows kept appearing; and the harder he tried, the more clearly they appeared.

Finally, exhausted, he went to the master and told him he was having terrible trouble carrying out his instructions. When the mahasiddha asked him what the problem was, the cowherd explained his difficulty.

"That's not really a problem," the master told him. "I can teach you another method. It's called cow meditation."

"What?" the cowherd asked, very surprised.

"I'm serious," the mahasiddha replied. "All you need to do is watch the images of the cows you see. Watch them as you lead them out to the pasture, as they graze, as you lead them back home to the farm. Whatever thoughts about cows appear to you, just watch them."

So the cowherd returned to his cave and sat down to practice with this new set of instructions. Because he wasn't trying to block his thoughts, this time his meditation proceeded very easily. He began to feel very peaceful and happy. He didn't miss his cows. And his mind grew calmer, more balanced, and more pliable.

After a while he went back to the mahasiddha and said, "Okay, now I've completed cow meditation. What do I do next?"

The master replied, "Very good. Now that you've learned to calm your mind, I'll teach you the second level of cow meditation. Here are the instructions: Meditate on your own body as a cow."

So the cowherd went back to his cave and began practicing as instructed, thinking, *Okay, now I'm a cow. I have horns and hooves, I make the sound* moo, *I eat grass. . . .* As he kept on with this practice, he found that his mind became even more peaceful and happy than before. When he felt he'd mastered this practice, he returned to the master and asked if there was a third level of instruction.

"Yes," the mahasiddha replied slowly. "For the third level of cow meditation, you have to focus on having horns."

So, once again, the cowherd returned to his cave to carry out his teacher's instructions, focusing only on the thought of having horns. He concentrated on the size of the horns, their placement, their color, the feeling of their weight on either side of his head. After a few months of practicing this way, he got up one morning and started to go outside to relieve himself. But as he tried to leave his cave, he felt something butting up against the sides of the cave's walls, making it impossible for him to get out. He reached up to figure out what the tacle was, and found to his surprise that two very long horns had routed from the sides of his head.

Turning his body sideways, he finally managed to get out of the cave, and ran, terrified, to his teacher.

"Look what happened!" he shouted. "You gave me this cow meditation, and now I've grown horns! This is horrible! It's like a nightmare!"

The mahasiddha laughed happily. "No, it's wonderful!" he exclaimed. "You've mastered the third level of cow meditation! Now you have to practice the fourth level. You have to think, 'Now I'm not a cow and I don't have horns.' "

Dutifully, the cowherd went back to his cave and practiced the fourth level of cow meditation, thinking, *Now I don't have horns, now I don't have horns, now I don't have horns. . . .* After a few days of practicing in this way, he woke up one morning and discovered that he could walk outside his cave without any difficulty at all. The horns had disappeared.

Surprised, he ran to the master, announcing, "Look, I don't have horns anymore! How does this happen? When I thought I had horns, they appeared. When I thought I didn't have horns, they disappeared. Why?"

The mahasiddha replied, "The horns came and went because of the way you focused your mind. The mind is very powerful. It can make experiences seem very real, and it can make them appear unreal."

"Oh," the cowherd said.

The master went on to explain, "Horns aren't the only things that appear and disappear according to the focus of your mind. Everything is like that. Your body, other people—the whole world. Their nature is emptiness. Nothing truly exists except in your mind's perception. Recognizing that is true vision. First you need to calm your mind, and then you learn how to see things clearly. This is the fifth level of cow meditation, learning to balance both tranquillity and true vision."

The cowherd returned once more to his cave, meditating with tranquillity and true vision. After a few years he became a mahasiddha himself, his mind having become calm and free from the circle of samsaric suffering.

There aren't many cowherds in the world anymore, though maybe it would be a more peaceful place if there were. Still, if you dare, you could practice like the old cowherd, but using a different object—like a car. After a few years of car meditation, you could end up a great master like the old cowherd. Of course, you'd have to be willing to spend a few years growing headlights, doors, seat belts, and maybe a trunk—and then learning how to make them disappear. And while you're practicing, you might find it hard to get in and out of your office elevator, and your coworkers might think it a little strange if you answered their questions with a honk instead of words.

I'm joking, of course. There are much easier ways of working with your thoughts than learning how to sprout cow horns or taillights.

USING YOUR THOUGHTS

When thoughts arise, instead of regarding them as faults,
recognize them to be empty and leave them just as they are.
                              —GÖTSANGPA, *The Highest Continuum,*
                              translated by Elizabeth M. Callahan

Even after you make friends with your five senses and learn to use sensory input as a support for meditation, you may find some difficulty in dealing with the "crazy monkey," the mental consciousness that enjoys leaping around, creating confusion, doubt, and uncertainty. Even if you learn to rest in simple sensory awareness, the crazy monkey mind will always be looking for new ways to interrupt whatever calmness, clarity, and openness you've achieved by offering a different and disturbing interpretation of events—a kind of psychological equivalent of throwing cushions around and gobbling up altar offerings. Difficult as it might be to deal with, the crazy monkey's interference is not a "bad" thing; it's simply a matter of entrenched neuronal patterns seeking to reassert themselves. The crazy monkey is essentially a neurologically programmed response to threats to human survival. Instead of getting angry, work with it. Why not generate a sense of gratitude toward its activity in helping us to survive?

Once you've learned to work with your senses, however, you need to deal with the crazy monkey itself, using the thoughts and emotions it generates as supports for calming the mind. And once you begin to work with these thoughts and emotions, you'll begin to discover a whole new dimension of freedom from ancient, survival-based patterns. You begin the process by questioning whether every thought you think, every feeling you feel, is a fact or a habit.

It's often the case that the first lessons we learn in life are the most important ones. "Look both ways before crossing the street." "Don't take candy from a stranger." "Don't play with matches." Children hear these things from their parents again and again, for good reason; and yet, as important as these childhood lessons are, we always seem to forget them. Human beings, by nature, take risks. That's how we learn. But some lessons can be deadly, while others can cause lasting pain. That's why, even as adults, we have to repeat the lessons we learned as children, and pass them on to our own children. Certain lessons just bear repeating.

So please forgive me if I reiterate something I learned early on in my formal training. *Thinking is the natural activity of the mind. Meditation is not about stopping your thoughts. Meditation is simply a process of resting the mind in its natural state, which is open to and naturally aware of thoughts, emotions, and sensations as they occur.* The mind is like a river, and, as with a river, there's no point in trying to stop its flow. You may as well try to stop your heart from beating or your lungs from breathing. What purpose would that serve?

But that doesn't mean you have to be a slave to whatever your mind produces. When you don't understand the nature and origin of your thoughts, your thoughts *use you.* When the Buddha recognized the nature of his mind, he reversed the process. He showed us how we can use our thoughts instead of being used *by* them.

When I first started training formally with my father, I was very nervous. I thought for sure he'd see how active my mind was and how many crazy thoughts leaped through it every second, and that he'd send me away because I wasn't a good candidate for learning. In one way, I was right. He did see how crazy my mind was. But I was wrong in thinking I was a bad candidate for meditation.

What he told me, along with his other students, was that no matter how many thoughts pass through your mind while you're meditating, it's okay. If a hundred thoughts pass through your mind in the space of a minute, you have a hundred supports for meditation. "How lucky you are!" he used to say. "If the crazy monkey inside your head is jumping all over the place, that's wonderful! Just watch the crazy monkey jumping around. Every jump, every thought, every distraction, like every sensory object, is a support for meditation. If you find yourself struggling with a lot of distractions, you can use every distraction as an object of meditation. Then they cease to be distractions and become supports for your meditation practice."

But he also warned us not to try to hold on to each thought as it arises. Whatever passes through the mind, we should just watch it come and go, lightly and without attachment, the way we'd practiced gently resting our attention on forms, sounds, or smells.

Watching thoughts is a bit like running to catch a bus. Just as you reach the bus stop, the bus is pulling away, so you have to wait for the next bus to come. In the same way, there's often a gap between thoughts—maybe it lasts for just a split second, but still, there's a gap. That gap is the experience of the complete openness of natural mind. Then another thought pops up, and when it disappears, there's another gap. Then another thought comes and goes, followed by another gap.

The process of observing your thoughts goes on and on in this way: thoughts followed by gaps, followed by thoughts, followed by gaps. If you continue this practice, very gradually the gaps become longer and longer, and your experience of resting the mind as it is becomes more direct. So there are two basic states of mind—with thought and without thought—and both are supports for meditation.

In the beginning, attention to thoughts always wavers. That's fine. If you find your mind wandering, just allow yourself to be aware of your mind wandering. Even daydreams can become the support for meditation if you allow your awareness to gently permeate them.

And when you suddenly remember, *Oops, I was supposed to be watching my thoughts, I was supposed to be focusing on form, I was supposed to be listening to sounds, I was supposed to be watching my thoughts,* just bring your attention back to whatever it was you were supposed to attend to. The great secret about these "Oops" moments is that they're actually split-second experiences of your fundamental nature.

It would be nice to hang on to every "Oops" you experience. But you can't. If you try, they harden into concepts—ideas about what "Oops" is supposed to mean. The good news is that the more you practice, the more "Oopses" you're likely to experience. And gradually these "Oopses" start to accumulate, until one day "Oops" becomes a natural state of mind, a release from habitual patterns of neuronal gossip that allows you to look at any thought, any feeling, and any situation with total freedom and openness.

"Oops" is a wonderful thing.

So now, try to practice "Oopsing" by bringing attention to your thoughts as supports for meditation. Just as in every other practice, it's important to start by just allowing your mind to rest in objectless awareness for a few moments, and then start watching your thoughts. Don't try to practice for very long. Give yourself a few minutes.

First, just rest your mind for a minute. . . .

Then let your mind become aware of your thoughts for maybe two minutes. . . .

And rest your mind again for a minute. . . .

When you're done, ask yourself what the experience was like for you. Did you have a lot of "Oopses"? Were you able to see your thoughts very clearly? Or were they hazy and indistinct? Or did they just vanish into thin air as soon as you tried to look at them?

When I teach this practice in public lectures, and afterward ask people about their experience, I get a lot of different answers. Some people say that when they try to watch their thoughts, the thoughts themselves become kind of sneaky. They vanish instantly or don't arise very clearly. Others say that their thoughts become very solid and clear, appearing in their minds as words, and they're able to watch thoughts come and go without much attachment or disturbance.

Now I'm going to let you in on a big secret: There is no secret! Both extremes that people describe—and anything in between—*are* experiences of meditation. If you're afraid of your thoughts, you're giving them power over you, because they seem so solid and real, so true.

And the more afraid of them you are, the more powerful they appear to be. But when you start to observe your thoughts, the power you give to them begins to fade. This can happen in one of two ways.

Sometimes, as mentioned earlier, if you watch your thoughts closely, you'll start to notice that they appear and disappear quite quickly, leaving little gaps between them. At first the gap between one thought and the next may not be very long; but with practice the gaps grow longer and your mind begins to rest more peacefully and openly in objectless meditation.

At other times, the simple practice of observing thoughts becomes something like watching TV or a movie. On the TV or movie screen, lots of things may be going on, but you are not actually *in* the movie or on the TV screen, are you? There's a little bit of space between yourself and whatever you're watching. When you practice observing your thoughts, you can actually experience that same sort of little bit of space between yourself and your thoughts. You're not really creating this space, because it was always there; you're merely allowing yourself to notice it. And through becoming aware of this space you can actually begin to enjoy watching your thoughts—even if they're scary—without being immersed in or controlled by them. Just let the thoughts spin out in their own way, the way adults watch children at play—building sand castles, fighting mock battles with plastic soldiers, or engaged in other games. The children are intensely involved in their activities, but adults simply watch and laugh affectionately at their seriousness.

Whichever of these experiences comes up for you is great, and no doubt the experiences will vary as you practice. Sometimes you'll observe your thoughts quite closely, seeing them come and go, noticing the gaps between them. Sometimes you'll simply watch them with that little bit of distance. Meditation is so much easier than most people think: Whatever you experience, as long as you are aware of what's going on, *is meditation!*

The only point at which your experience shifts from meditation into something else occurs when you try to control or change whatever you might be experiencing. But if you bring some awareness to your attempt to control your experience, that's meditation, too.

Of course, some people don't see any thoughts at all; their minds just go blank. That's fine, too. It's *your* mind you're working with, so no one can judge you; no one can grade you on your experience. Meditation is a uniquely personal process, and no two people's experiences are alike. As you continue practicing, you'll undoubtedly find that your own experience shifts sometimes from day to day, from practice period to practice period. Sometimes you may find your thoughts are very clear and easy to observe; sometimes they may seem quite vague and slippery. Sometimes you may find that your mind becomes dull or foggy when you sit down to practice. That's okay, too. The sense of dullness is little more than a chain of neurons chattering with one another in response to your intention to meditate, and you can simply observe the dullness or whatever else you may feel. Observation, giving bare attention to whatever you happen to be experiencing at a particular moment, *is* meditation. Even the neuronal gossip that manifests as a thought like "I don't know how to meditate" can be a support for meditation as long as you observe it.

As long as you maintain awareness or mindfulness, no matter what happens when you practice, your practice is meditation. If you watch your thoughts, that is meditation. If you can't watch your thoughts, that is meditation, too. Any of these experiences can be supports for meditation. The essential thing is to maintain awareness, no matter what thoughts, emotions, or sensations occur. If you remember that awareness of whatever occurs *is* meditation, then meditation becomes much easier than you may think.

THE SPECIAL CASE OF UNPLEASANT THOUGHTS

> *No matter what thought occurs, don't try to stop it.*
>> —THE NINTH GYALWANG KARMAPA,
>> *Mahāmudrā: The Ocean of Definitive Meaning*,
>> translated by Elizabeth M. Callahan

Especially if you're new to meditation, it can be very difficult to observe thoughts related to unpleasant experiences—particularly those aligned with strong emotions such as jealousy, anger, fear, or envy—with bare attention. Such unpleasant thoughts can be so strong and persistent that it's easy to get caught up in following after them. I don't have enough fingers and toes to count the number of people I've met who've discussed this problem with me, especially if the thoughts they're experiencing relate to fights they've had with someone at home, in the office, or some other place that they can't forget. Day after day, their minds keep going back to the ideas they attach to what was said and done, and they find themselves caught up in thinking about how terrible the other person was, what they could or should have said at the time, and what they'd like to do to get revenge.

The best way to work with these kinds of thoughts is to step back and rest your mind in objectless shinay for a minute, and then bring your attention to each thought and the ideas that revolve around it, observing both directly for a few minutes, just as you would observe the shape or color of a form. Allow yourself to alternate between resting your mind in objectless meditation and bringing your attention back to the same thoughts.

When you work with negative thoughts in this way, two things happen. (Don't worry—neither one of them involves growing horns!) First, as you rest in awareness, your mind begins to settle. Second, you'll find that your attention to particular thoughts or stories comes and goes, just the way it does when working with forms, sounds, and other sensory supports. And as that thought or story is interrupted by other issues—like folding the laundry, buying groceries, or preparing for a meeting—the unpleasant ideas gradually lose their grip on your mind. You begin to realize that they're not as solid or powerful as they first appeared. It's more like a busy signal on the telephone—annoying, perhaps, but nothing you can't deal with.

When you work with unpleasant thoughts in this way, they become assets to mental stability rather than liabilities—like adding weight to the bar when you're exercising in a gym. You're developing psychological muscles to cope with greater and greater levels of stress.

USING EMOTIONS

One does not have to feel totally at the mercy of one's emotion.
—KALU RINPOCHE, *Gently Whispered,*
edited by Elizabeth Selanda

Because emotions tend to be vivid and enduring, they can be even more useful than thoughts as supports for meditation. My father and my other teachers impressed on me that there are three basic categories of emotion: positive, negative, and neutral.

Positive emotions—such as love, compassion, friendship, and loyalty—strengthen the mind, build our confidence, and enhance our ability to assist those in need of help. Some translations of Buddhist texts refer to such emotions, and the actions connected with them, as "virtuous"—a translation that, at least as I've observed among Western audiences, seems to have some sort of moral association. In fact, there is no moral or ethical association connected with such actions and emotions. As explained to me by a student who has some knowledge of the meanings of Western words, the word *virtue,* as applied to the translation of the Tibetan term *gewa,* is more closely related to the ancient meaning of "virtue" as potency or effectiveness in terms of healing power.

Negative emotions, such as fear, anger, sadness, jealousy, grief, or envy—often translated as "nonvirtuous" (or, in Tibetan, *mi-gewa*) feelings—are emotions that tend to weaken the mind, undermine confidence, and increase fear.

More or less neutral feelings, meanwhile, basically consist of ambivalent responses—the kinds of feelings we might have toward a pencil, a piece of paper, or a staple remover. Try as you might, it's hard to feel positively or negatively toward a pencil!

The method of using emotions as supports differs depending on the type of emotion you're experiencing. If you're feeling a positive emotion, the type that strengthens your mind, you can focus on both the feeling *and* the object of the feeling. For example, if you're feeling love for a child, you can rest your attention on both the child *and* the love you feel for him or her. If you're feeling compassion for someone in trouble, you can focus on the person needing help *and* your feeling of compassion. In this way, the object of your emotion becomes a support for the emotion itself, while the emotion becomes a support for focusing on the object that inspires the emotion.

Conversely, holding an object of negative emotion in attention tends to reinforce a mental image of that person, situation, or thing as something *bad in itself*. No matter how much you try to cultivate compassion, confidence, or any other positive feeling, your mind will almost automatically associate the object with the negative emotion: "Whoa, that one is *bad*. Fight it. Make it go away. Or run away."

A more constructive approach to negative emotions, similar to working with negative thoughts, is simply to rest your attention on the emotion itself rather than on its object. Just look at the emotion without analyzing it intellectually. Don't try to hold on to it and don't try to block it. Just observe it. When you do this, the emotion won't seem as big or powerful as it initially did.

This is the same sort of process I practiced during my first year of retreat, when the fear and anxiety I felt around other people forced me to run back to sit alone in my own room. Once I began to simply observe my fears, I began to see that they weren't solid, indivisible monsters that I could never overcome, but instead a series of small, fleeting sensations and images that popped in and out of awareness so rapidly that they only gave the semblance of being solid and whole (similar, as I would later discover, to the way a whirling mass of subatomic particles produces the appearance of something indivisible and solid). And after observing my fear this way, I started to think, *Hmm, that's interesting. This fear isn't so big and powerful at all. In fact, it's pretty harmless. It's just a bunch of transitory sensations that appear, hang around for a second or two, and then simply disappear.*

This didn't happen overnight, of course. I had to spend a few weeks completely immersed in the process, like some sort of mad scientist utterly absorbed in an experiment. I also had the benefit of several years of training to support me.

But I emerged from the experience with a new appreciation for all the different methods the Buddha had provided so many centuries ago to assist people he would never personally know in overcoming such difficulties. Later, when I began to learn more about the structure and function of the brain and the insights into the nature of reality described by modern physicists, I was even more impressed by the parallels between the techniques the Buddha had arrived at through introspection and the explanations achieved through objective observation as to why they worked.

Sometimes, though, the object associated with a negative emotion—whether it's a person, a place, or an event—is just too clear or present to ignore. If that's the case, by all means don't try to block it. Use it. Rest your attention on the form, smell, taste, or any of the other sensory perceptions you learned to work with earlier on. In this way, the object of the emotion can become, in itself, a very powerful support for meditation.

This approach is useful when you begin to work directly with the basic mental afflictions described in Part One of this book. When I was introduced to the subject of mental afflictions, I thought, *Oh no, I'm flawed. I'm ignorant. I have a lot of attachments and aversions. I'm stuck with unhappiness for the rest of my life.* But then I heard an old proverb. I don't know if it's based on fact, but it goes something like this: "Peacocks eat poison, and the poison they eat is transformed into beautiful feathers."

Having spent most of my early life crushed into a little ball of fear and anxiety, I know how strong mental afflictions can be. I spent thirteen years thinking I would die—and sometimes hoping I would, just to be free of the fear I felt. It wasn't until I entered retreat and had to face these afflictions head-on that I learned that ignorance, attachment, and aversion were the material I was given to work with, which, like the poison peacocks eat, turned out to be the source of great blessing.

Every mental affliction is actually the basis of wisdom. If we get caught up in our afflictions or try to repress them, we just end up creating more problems for ourselves. If, instead, we look at them directly, the things we fear will kill us are gradually transformed into the strongest supports for meditation we could ever hope for.

Mental afflictions are not enemies. They're our friends.

That's a hard truth to accept. But every time you recoil from it, think of the peacock. Poison doesn't taste very good. But if you swallow it, it turns into beauty.

So, in our final lesson in practice, we're going to look at meditative antidotes we can apply when facing our most fearsome and unpleasant experiences. As we examine these practices, we'll come to recognize that the degree to which any experience repels, frightens, or seems to weaken us is equal to the degree to which such experiences can make us stronger, more confident, more open, and more able to accept the infinite possibilities of our Buddha nature.

13

COMPASSION: OPENING THE HEART OF THE MIND

Look on everyone you see with an open, loving heart.
—ŚĀNTIDEVA, *The Bodhicaryāvatāra*
translated by Kate Crosby and Andrew Skilton

BECAUSE WE ALL live in a human society on a single planet, we have to learn to work together. In a world bereft of compassion, the only way we can work together is through the enforcement of outside agencies: police, armies, and the laws and weapons to back them up. But if we could learn to develop loving-kindness and compassion toward one another—a spontaneous understanding that whatever we do to benefit ourselves must benefit others and vice versa—we wouldn't need laws or armies, police, guns, or bombs. In other words, the best form of security we can offer ourselves is to develop an open heart.

I've heard some people say that if everyone was kind and compassionate, the world would be a boring place. People would be no more than sheep, idling around with nothing to do. Nothing could be further from the truth. A compassionate mind is a diligent mind. There's no end of problems in this world: Thousands of children die every day from starvation; people are slaughtered in wars that never even get reported in the newspapers; poisonous gases are building up in the atmosphere, threatening our very existence. But we don't even have to look so far and wide to find suffering. We can see it all around us: in coworkers going through the pain of divorce; in relatives coping with physical or mental illness; in friends who've lost their jobs; and in hundreds of animals put to death every day because they're unwanted, lost, or abandoned.

If you really want to see how active a compassionate mind can be, here's a very simple exercise that probably won't take more than five minutes of your time. Sit down with a pen and paper and make a list of ten problems that you'd like to see solved. It doesn't matter whether they're global problems or issues close to home. You don't have to come up with solutions. Just write the list.

The simple act of writing this list will change your attitude significantly. It will awaken the natural compassion of your own true mind.

THE MEANING OF LOVING-KINDNESS AND COMPASSION

If we were to make a list of people we don't like . . . we would
find a lot about those aspects of ourselves that we can't face.
—PEMA CHÖDRON, *Start Where You Are*

Recently, a student of mine told me that he thought "loving-kindness" and "compassion" were cold terms. They sounded too distant and academic, too much like an intellectual exercise in feeling sorry for other people. "Why," he asked, "can't we use a simpler, more direct word, like 'love'?"

There are a couple of good reasons why Buddhists use the terms "loving-kindness" and "compassion" instead of a simpler one like "love." Love, as a word, is so closely connected with the mental, emotional, and physical responses associated with desire that there's some danger in associating this aspect of opening the mind with reinforcing the essentially dualistic delusion of self and other: "I love *you*," or "I love *that*." There's a sense of dependence on the beloved object, and an emphasis on one's personal benefit in loving and being loved. Of course, there are examples of love, such as the connection between a parent and a child, that transcend personal benefit to include the desire to benefit another. Most parents would probably agree that the love they experience toward their children involves more sacrifice than personal reward.

By and large, though, the terms "loving-kindness" and "compassion" serve as linguistic "stop signs." They make us pause and think about our relationship to others. From a Buddhist perspective, loving-kindness is the aspiration that all other sentient beings—even those we dislike—experience the same sense of joy and freedom that we ourselves aspire to feel: a recognition that we all experience the same *kinds* of wants and needs; the desire to go about our lives peacefully and without fear of pain or harm. Even an ant or a cockroach experiences the same *kinds* of needs and fears that humans do. As sentient beings, we're all alike; we're all kindred. Loving-kindness implies a sort of challenge to develop this awareness of kindness or commonality on an emotional, even physical, level, rather than allowing it to remain an intellectual concept.

Compassion takes this capacity to look at another sentient being as equal to oneself even further. Its basic meaning is "feeling with," a recognition that what you feel, I feel. Anything that hurts you hurts me. Anything that helps you helps me. Compassion, in Buddhist terms, is a complete identification with others and an active readiness to help them in any way.

Just look at it practically. If you lie to someone, for example, who have you really hurt? *Yourself.* You have to bear the burden of remembering the lie you told, covering your tracks, and maybe spinning a whole web of new lies to keep the original lie from being discovered. Or suppose you steal something, even something as small as a pen, from your office or some other place. Just think about the number of large and small actions you have to take to hide what you did. And despite all the energy you put into concealing what you did, you'll almost inevitably be caught. There's no way you can hide every single detail. So, in the end, all you've really done is wasted a lot of time and effort that you might have spent doing something more constructive.

Compassion is essentially the recognition that everyone and everything is a reflection of everyone and everything else. An ancient text called the *Avatamsaka Sutra* describes the universe as an infinite net brought into existence through the will of the Hindu god Indra. At every connection in this infinite net hangs a magnificently polished and infinitely faceted jewel, which reflects in each of its facets all the facets of every other jewel in the net. Since the net itself, the number of jewels, and the facets of every jewel are infinite, the number of reflections is infinite as well. When any jewel in this infinite net is altered in any way, all of the other jewels in the net change, too.

The story of Indra's net is a poetic explanation for the sometimes mysterious connections that we observe between seemingly unrelated events. I've heard from a number of students recently that a lot of modern scientists have been grappling for a long time with the question of connections—or entanglements, as they are known to physicists—between particles that aren't readily obvious to the human mind or to a microscope. Apparently, experiments involving subatomic particles conducted over the past few decades suggest that anything that was connected at one time retains that connection forever. Like the jewels of Indra's net, anything that affects one of these tiny particles automatically affects another, regardless of how far they're separated by time or space. And since one of the current theories of modern physics holds that all matter was connected as a single point at the start of the big bang that created our universe, it's *theoretically* possible—though as yet unproven—that whatever affects one particle in our universe also affects every other one.

The profound interrelatedness suggested by the story of Indra's net, though currently only an analogy to contemporary scientific theory, may one day turn out to be scientific fact. And that possibility, in turn, transforms the whole idea of cultivating compassion from a nice idea into a matter of life-shaking proportions. Just by changing your perspective, you can not only alter your own experience, you can also change the world.

TAKING IT SLOWLY

> *Be free from any grasping for experience.*
>
> —The Ninth Gyalwang Karmapa,
> *Mahāmudrā: The Ocean of Definitive Meaning,*
> translated by Elizabeth M. Callahan

Training in loving-kindness and compassion has to be undertaken gradually. It's too easy, otherwise, to take on too much, too soon—a tendency exemplified by a cautionary tale I was told when I began this phase of my training. The story concerns Milarepa, widely regarded as one of Tibet's greatest enlightened masters, who taught mainly through songs and poems he composed on the spot. During his lifetime, Milarepa traveled a great deal, and one day he arrived in a village and sat down to sing. One of the villagers heard his song and became completely entranced by the idea of giving up everything he was attached to and living as a hermit, in order to become enlightened as fast as possible and to help as many people in the world as he could during the years remaining to him.

When he told Milarepa what he meant to do, Milarepa gently advised him that it might be a better idea to stay at home for a while and start practicing compassion in a more gradual way. But the man insisted that he wanted to abandon everything right away, and, ignoring Milarepa's advice, he rushed home and feverishly began giving away everything he owned, including his house. After tying a few necessary items in a handkerchief, he left for the mountains, found a cave, and sat down to meditate, without ever having practiced before and without ever having taken the time to learn how. Three days later, though, the poor man was hungry, exhausted, and freezing. After five days of starvation and discomfort, he wanted to go home, but was too embarrassed to do so. *I made such a show about leaving everything and going to meditate,* he thought, *what will people think if I come back after only five days?*

By the end of the seventh day, though, he couldn't take the cold and hunger any longer and returned to his village. Sheepishly, he went around to all his neighbors asking if they would mind giving back his things. They returned everything he'd given away, and after he'd resettled himself, the man went back to Milarepa and, thoroughly humbled, asked for preliminary instructions in meditation. Following the gradual path Milarepa taught him, he eventually became a meditator of great wisdom and compassion and was able to benefit many others.

The moral of the story, of course, is to resist the temptation to rush into practice expecting immediate results. Since our dualistic perspective of "self" and "other" didn't develop overnight, we can't expect to overcome it all at once. If we rush onto the path of compassion, at best we'll end up like the villager who rashly gave up everything he owned. At worst we'll end up regretting a charitable act, creating for ourselves a mental obstacle that may take years to overcome.

This point was repeatedly impressed on me by my father and my other teachers. If you take a gradual path, your life might not change tomorrow, next week, or even a month from now. But as you look back over the course of a year or many years, you *will* see a difference. You'll find yourself surrounded by loving and supportive companions. When you come into conflict with other people, their words and actions won't seem as threatening as they once did. Whatever pain or suffering you might sometimes feel will assume much more manageable, life-sized proportions, maybe even shrinking in importance compared with what other people you know may be going through.

The gradual path I was taught regarding the development of compassion toward others consisted of three "levels," each to be practiced for several months—similar to the way students learn basic math skills—before moving on to higher applications. "Level One" involved learning how to develop a kind or compassionate attitude toward yourself and other beings close to you. "Level Two" focused on developing immeasurable loving-kindness and compassion toward all beings. "Level Three" involves cultivating bodhicitta.

There are actually two types, or levels, of bodhicitta: absolute and relative. *Absolute bodhicitta* is a spontaneous recognition that all sentient beings, regardless of how they act or appear, are already completely enlightened. It usually takes a good deal of practice to attain this level of spontaneous recognition. *Relative bodhicitta* involves the cultivation of the desire for all sentient beings to become completely free of suffering through recognizing their true nature, and taking the actions to accomplish that desire.

LEVEL ONE

When you think of a condemned prisoner . . . imagine it's you.
—Patrul Rinpoche, *The Words of My Perfect Teacher,*
translated by the Padmakara Translation Group

Meditation on loving-kindness and compassion shares many similarities with the shinay practices we've already talked about. The main difference is the choice of the object on which we rest our attention and the methods we use to rest our attention. One of the most important lessons I learned during my years of formal training was that whenever I blocked the compassion that is a natural quality of my mind, I inevitably found myself feeling small, vulnerable, and afraid.

It's so easy to think that *we're* the only ones who suffer, while other people are somehow immune to pain, as though they'd been born with some kind of special knowledge about being happy that, through some cosmic accident, we never received. Thinking in this way, we make our own problems seem much bigger than they really are.

I've been as guilty of this belief as anyone else, and, as a result, have allowed myself to become isolated, trapped in a dualistic mode of thinking, pitting my weak, vulnerable, and fearful self against everyone else in the world, whom I thought of as much more powerful, happy, and secure. The power I fooled myself into believing other people held over me became a terrible threat to my own well-being. At any given moment, I thought, someone could find a way to undermine whatever security or happiness I'd managed to achieve.

After working with people over the years, I've come to realize that I wasn't the only person to feel this way. Some part of our ancient, reptilian brain immediately evaluates whether we're facing a friend or an enemy. This perception gradually extends even to inanimate objects until everything—a computer, a blown fuse, the blinking light on an answering machine—seems somewhat menacing.

When I began to practice meditation on compassion, however, I found that my sense of isolation began to diminish, while at the same time my personal sense of empowerment began to grow. Where once I saw only problems, I started to see solutions. Where once I viewed my own happiness as more important than the happiness of others, I began to see the well-being of others as the foundation of my own peace of mind.

The way I was taught, the development of loving-kindness and compassion begins with learning how to appreciate oneself. This is a hard lesson, especially for people brought up in cultures in which it's common to dwell on personal weaknesses rather than on personal strengths. This is not a particularly Western problem. Developing a compassionate attitude toward myself literally saved my life during my first year in retreat. I could never have left my room if I hadn't come to terms with my real nature, looking deep within my own mind and seeing the real power there, instead of the vulnerability I'd always thought was there.

One of the things that helped me as I sat alone in my room was remembering that the Sanskrit word for "human being" is *purusha,* which basically means "something that possesses power." Being human means having power; specifically, the power to accomplish whatever we want. And what we want goes back to the basic biological urge to be happy and to avoid pain.

So, in the beginning, developing loving-kindness and compassion means using yourself as the object of your meditative focus. The easiest method is a kind of variation on the "scanning practice" described earlier. If you're practicing formally, assume the seven-point posture to the best of your ability. Otherwise, just straighten your spine while keeping the rest of your body relaxed and balanced, and allow your mind simply to relax in a state of bare awareness.

After a few moments of resting your mind in objectless meditation, do a quick "scanning exercise," gradually observing your physical body. As you scan your body, gently allow yourself to recognize how wonderful it is just to *have* a body, as well as a mind that's capable of scanning it. Allow yourself to recognize how magnificent these very basic facts of your existence really are, how fortunate you are simply to have the great gifts of a body and a mind! Rest in that recognition for a moment, and then gently introduce the thought *How nice it would be if I were always able to enjoy this sense of basic goodness. How nice it would be if I could always enjoy this sense of well-being and all the causes that lead to feeling happy, peaceful, and good.*

Then just allow your mind to rest, open and relaxed. Don't try to keep up this practice for more than three minutes if you're practicing formally, or for more than a few seconds during informal meditation sessions. It's very important to practice in short sessions and then allow your mind to rest. Short practice sessions followed by periods of rest allow this new awareness to stabilize—or, in Western scientific terms, give your brain a chance to establish new patterns without being overwhelmed by old neuronal gossip. Very simply, when you let go of practicing, you give yourself a chance to let the effects simply wash over you in a flood of positive feeling.

Once you've become somewhat familiar with your own desire for happiness, extending that awareness to other sentient beings around you—people, animals, and even insects—becomes much easier. The practice of loving-kindness and compassion toward others essentially involves cultivating the recognition that all living creatures want to feel whole, safe, and happy. All you have to do is remember that whatever's going on inside someone else's mind is the same thing that's going on in yours. When you remember this, you realize that there's no reason to be frightened of anyone or anything. The only reason you're ever scared is that you've failed to recognize that whomever or whatever you're facing is just like you: a creature that only wants to be happy and free from suffering.

The classic Buddhist texts teach that we should focus first on our mothers, who have shown the ultimate kindness toward us by carrying us in their bodies, bringing us into the world, and nurturing us throughout the early years of our lives, often at great sacrifice to themselves. I understand that many people in Western cultures don't always enjoy tender and affectionate relationships with their parents—in which case using one's mother or father as an object of meditation wouldn't be very practical. It's perfectly acceptable in such cases to focus on another object, such as an especially kind relative, a teacher, a close friend, or a child. Some people choose to focus on their pets. The object of your meditation doesn't really matter; the important thing is to rest your attention lightly on someone or something toward which you feel a deep sense of tenderness or warmth.

When you take up loving-kindness and compassion as a formal practice, begin by assuming either the seven-point posture or, at the very least (if you're sitting on a bus or a train, for example), straightening your spine while allowing the rest of your body to rest naturally. As with any meditation practice, once you've got your body positioned, the next step is simply to allow your mind to rest naturally for a few moments and let go of whatever you might have been thinking about. Just let your mind breathe a huge sigh of relief.

After resting your mind for a few moments in objectless meditation, lightly bring your awareness to someone toward whom it's easiest for you to feel tenderness, affection, or concern. Don't be surprised if the image of someone or something you didn't deliberately choose appears more strongly than the object you may have decided to work with. This happens, often quite spontaneously. One of my students began formal practice intending to focus on his grandmother, who had been very kind to him when he was young; but the image that kept appearing to him was a rabbit he'd owned as a child. This is just an example of the mind's natural wisdom asserting itself. He actually had a lot of warm memories associated with the rabbit, and when he finally surrendered to them, his practice became quite easy.

Sometimes you may find that your mind spontaneously produces memories of a particularly nice experience you shared with someone, rather than a more abstract image of the person you've chosen as an object of meditation. That's fine, too. The important point in cultivating loving-kindness and compassion is to allow yourself to experience genuine feelings of warmth, tenderness, or affection.

As you proceed, allow the sense of warmth or affection to settle in your mind, like a seed planted in soil, alternating for a few minutes between this experience and simply allowing your mind to rest in objectless meditation. As you alternate between these two states, allow yourself to wish that the object of your meditation might experience the same sense of openness and warmth you feel toward him or her.

After practicing in this way for a while, you're ready to move a little bit deeper. Begin as before, assuming an appropriate posture and allowing your mind to rest in objectless meditation for a few moments, and then bring to mind the object of your loving-kindness and compassion. Once you've settled on the object of your meditation, there are a couple of ways to proceed. The first is to imagine the object you've chosen in a very sad or painful state. Of course, if the object you've chosen is already in deep pain or sorrow, you can simply bring to mind his or her present condition. Either way, the image you call to mind naturally produces a profound sense of love and connectedness, and a deep desire to help. Thinking that someone or something you care for is in pain can break your heart. But a broken heart is an open heart. Every heartbreak is an opportunity for love and compassion to flow through you.

Another approach is to rest your attention lightly on the subject you've chosen while asking yourself, "How much do *I* want to be happy? How much do *I* want to avoid pain or suffering?" Let your thoughts on these points be as specific as possible. For example, if you're stuck somewhere stiflingly hot, would you rather move to a cooler and more open place? If you feel some sort of physical pain, would you like the pain to be lifted? As you think about your own answers, gradually turn your attention to the subject you've chosen and imagine how he or she would feel in the same situation. Practicing in this way not only opens your heart to other beings, but also dissolves your own identification with whatever pain or discomfort you may be experiencing at the moment.

Cultivating loving-kindness and compassion toward those you know and care about already isn't so hard because even when you want to strangle them for being stupid or obstinate, the bottom line is that you still love them. It's a little bit harder to extend the same sense of warmth and relatedness toward those you don't know—and even harder to extend that awareness to those you actively dislike.

I heard a story a while back about a man and a woman living in China, maybe forty or fifty years ago. They'd just gotten married, and when the bride moved into her husband's home, she immediately started fighting with her mother-in-law over a number of petty issues about how the household should be run. Gradually their disagreements escalated until the new bride and her mother-in-law couldn't even stand to look at each other. The bride saw her mother-in-law as an interfering old witch, while the mother-in-law thought of her son's young bride as an arrogant child, with no respect for her elders.

There was no real reason for their anger to escalate as it had. But eventually the bride became so angry at her mother-in-law that she decided she had to do something to get her out of the way. So she went to a doctor and asked for poison that she could put into her mother-in-law's food.

After hearing the young bride's complaints, the doctor agreed to sell her some poison. "But," he warned, "if I were to give you something strong that worked immediately, everyone would point their fingers at you and say, 'You poisoned your mother-in-law,' and they'd also find out that you bought the poison from me, and that wouldn't be good for either of us. So I'm going to give you a gentle poison that will take effect very gradually, so she won't die right away."

He also instructed her that while she administered the poison, she should treat her mother-in-law very, very nicely. "Serve every meal with a smile," he advised. "Tell her you hope she enjoys her food and ask her if there's anything else you might bring her. Be very humble and sweet, so no one will suspect you."

The bride agreed, and carried the poison back home with her. That very evening she started adding the poison to her mother-in-law's food, and very politely offered the meal to her. After a few days of being treated so respectfully, the mother-in-law began to change her opinion about her son's wife. *Maybe she's not so arrogant after all*, the old woman thought. *Maybe I was wrong about her.* And little by little she started treating her daughter-in-law more agreeably, complimenting her on her cooking and on the way she was managing the household, and even exchanging little tidbits of gossip and funny stories.

As the old woman's attitude and behavior changed, of course, so did the girl's. After a few days she started thinking, *Maybe my mother-in-law isn't as bad as I thought. In fact, she seems pretty nice.*

This continued for a month or so, until the two women actually started to become very good friends. They started getting along so well that at a certain point the girl stopped poisoning her mother-in-law's food. Then she started to worry because she realized she'd already put so much poison in every meal that her mother-in-law would very likely die.

So she went back to the doctor and told him, "I made a mistake. My mother-in-law is actually a really nice person. I shouldn't have poisoned her. Please help me out and give me an antidote to the poison I gave her."

The doctor sat very quietly for a moment after listening to the girl. "I'm very sorry," he told her. "I can't help you now. There is no antidote."

On hearing that, the girl became terribly upset and started to cry, swearing that she was going to kill herself.

"Why would you want to kill yourself?" the doctor asked.

The girl answered, "Because I've poisoned such a nice person and now she's going to die. So I should take my own life to punish myself for the terrible thing I did."

Again the doctor sat quietly for a moment, and then he started to chuckle.

"How can you laugh about this?" the girl demanded.

"Because there's really no need for you to worry," he replied. "There's no antidote to the poison because I never gave you any poison to begin with. All I gave you was a harmless herb."

I like this story because it's such a simple example of how easily a natural transformation of experience can occur. At first the new bride and her mother-in-law hated each other. Each thought that the other was simply awful. Once they started to treat each other differently, though, they began to see each other in a different light. Each saw the other as a basically good person, and eventually they became close friends. As people, they hadn't really changed at all. The only thing that had changed was their perspective.

The nice thing about such stories is that they compel us to see that our initial impressions about others may be wrong or misguided. There's no reason to feel guilty about such mistakes; they're merely the result of ignorance. And, fortunately, the Buddha provided a meditation practice that provides not only the means for amending such mistakes, but for preventing them in the future. This practice is known in

English as "exchanging yourself for others," which in simple terms means imagining yourself in the position of someone or something you don't like very much.

Although the practice of exchanging yourself for others can be performed anytime, anywhere, it's helpful to get the basics down through formal practice. Formal practice is a bit like charging the battery in a cell phone. Once the battery is fully charged, you can use the cell phone for a long time, in a variety of places and under a variety of circumstances. Eventually, though, the battery runs down, and you need to charge it again. The main difference between charging a battery and developing loving-kindness and compassion is that ultimately, through formal practice, the habit of responding compassionately to other beings creates a series of neuronal connections that constantly perpetuates itself and doesn't lose its "charge."

The first step in formal practice is, as usual, to assume a correct posture and allow your mind to rest for a few moments. Then bring to mind someone or something that you *don't* like. Don't judge what you feel. Give yourself complete permission to feel it. Simply letting go of judgments and justifications will let you experience a certain degree of openness and clarity.

The next step is to admit to yourself that whatever you're feeling—anger, resentment, jealousy, or desire—is in itself the source of whatever pain or discomfort you're experiencing. The object of your feeling isn't the source of your pain, but rather your own mentally generated response to whomever or whatever you're focusing on.

For example, you might bring your attention to someone who's said something to you that sounded cruel, critical, or contemptuous—or even to someone who has told you an outright lie. Then, allow yourself to recognize that all that has occurred is that someone has emitted sounds and you have heard them. If you've spent even a little bit of time practicing calm-abiding meditation on sound, this aspect of "exchanging self for others" will probably feel familiar.

At this point, three options are available to you. The first, and most likely, option is to allow yourself to be consumed by anger, guilt, or resentment.

The second (which is very unlikely) is to think, *I should have spent more time meditating on sound.*

The third option is to imagine yourself as the person who said or did whatever you felt as painful. Ask yourself whether what that person said or did was really motivated by a desire to hurt you, or whether he or she was trying to alleviate his or her own pain or fear.

In many cases, you know the answer already. You may have over-heard some talk about the other person's health or relationship, or some threat to his or her professional standing. But even if you don't know the specifics of a person's situation, you'll know from your own practice of developing compassion for yourself and of extending it toward others that there is only one possible motive behind someone's behavior: the desire to feel safe or happy. And if people say or do something hurtful, it's because they don't feel safe or happy. In other words, they're scared.

And you know what it's like to be scared.

Recognizing this about someone else is the essence of exchanging self for others.

Another method of exchanging yourself for others is to choose a "neutral" focus—a person or an animal you may not know directly, but whose suffering you're somewhat aware of. Your focus could be a child in a foreign country, dying of thirst or hunger, or an animal caught in a steel trap, desperately chewing off its leg to escape. These "neutral" beings experience all kinds of suffering over which they have no control and from which they cannot protect or free themselves. Yet the pain they feel and their desperate desire to free themselves from it are easily understandable, because you share the same basic longing. So, even though you don't know them, you recognize their state of mind, and experience their pain and fear as your own. I'm willing to bet that extending compassion in this way—toward those you don't like or those you don't know—won't turn you into a boring, lazy old sheep.

LEVEL TWO

May all beings have happiness and the causes of happiness.
                                        —*The Four Immeasurables*

There's a particular meditation practice that can help generate immeasurable loving-kindness and compassion. In Tibetan, this practice is called *tonglen,* which may be translated into English as "sending and taking."

Tonglen is actually quite an easy practice, requiring only a simple coordination of imagination and breathing. The first step is simply to recognize that as much as you want to achieve happiness and avoid suffering, other beings also feel the same way. There is no need to visualize specific beings, although you may start out with a specific visualization if you find it helpful. Eventually, however, the practice of taking and sending extends beyond those you can imagine to include all sentient beings—including animals, insects, and inhabitants of dimensions you don't possess the knowledge or capacity to see.

The point, as I was taught, is simply to remember that the universe is filled with an infinite number of beings, and to think, *Just as I want happiness, all beings want happiness. Just as I wish to avoid suffering, all beings wish to avoid suffering. I am just one person, while the number of other beings is infinite. The well-being of this infinite number is more important than that of one.* And as you allow these thoughts to roll around in your mind, you'll actually begin to find yourself actively engaged in wishing for others' freedom from suffering.

Begin by assuming a correct posture and allowing your mind to simply rest for a few moments. Then use your breath to send all your happiness to all sentient beings and absorb their suffering. As you exhale, imagine all the happiness and benefits you've acquired during your life pouring out of yourself in the form of pure light that spreads to all beings and dissolves into them, fulfilling all their needs and eliminating their suffering. As soon as you start to breathe out, imagine the light immediately touching all beings, and that by the time you finish exhaling, the light has already dissolved into them. As you inhale, imagine the pain and suffering of all sentient beings as a dark, smoky light being absorbed through your nostrils and dissolving into your heart.

As you continue this practice, imagine that all beings are freed from suffering, and filled with bliss and happiness. After practicing in this way for a few moments, simply allow your mind to rest. Then take up the practice again, alternating between periods of tonglen and resting your mind.

If it helps your visualization, you can sit with your body very straight and rest your hands in loosely closed fists on the tops of your thighs. As you breathe out, open your fingers and slide your hands down your thighs toward your knees while you imagine the light going out toward all beings. As you inhale, slide your hands back up, forming loosely closed fists as through drawing the dark light of others' suffering and dissolving it into yourself.

The universe is filled with so many different kinds of creatures, it's impossible even to imagine them all, much less offer direct and immediate help to each and every one. But through the practice of tonglen, you open your mind to infinite creatures and wish for their well-being. The result is that eventually your mind becomes clearer, calmer, more focused and aware, and you develop the capacity to help others in infinite ways, both directly and indirectly.

An old Tibetan folktale illustrates the benefits of developing this sort of all-encompassing compassion. A nomad who spent his days walking across the mountains was constantly pained by the rough and thorny ground because he didn't have any shoes. Over the course of his travels, he began to collect the skins of dead animals and spread them along the mountain paths, covering the stones and thorns. The problem was that even with great effort, he could only cover several hundred square yards. At last it came to him that if he simply used a few small hides to make himself a pair of shoes, he could walk for thousands of miles without any pain. Simply by covering his feet with leather, he covered the entire earth with leather.

In the same way, if you try to deal with each conflict, each emotion, and each negative thought as it occurs, you're like the nomad trying to cover the world with leather. If, instead, you work at developing a loving and peaceful mind, you can apply the same solution to every problem in your life.

LEVEL THREE

A person who has . . . awakened the force of genuine compassion will be quite capable of working physically, verbally, and mentally for the welfare of others.

—Jamgön Kongtrul, *The Torch of Certainty,*
translated by Judith Hanson

The practice of bodhicitta—the mind of awakening—may seem almost magical, in the sense that when you choose to deal with other people as if they were already fully enlightened, they tend to respond in a more positive, confident, and peaceful manner than they otherwise might. But really there is nothing magical about the process. You're simply looking at and acting toward people on the level of their full potential, and they respond to the best of their ability in the same way.

As mentioned earlier, there are two aspects of bodhicitta, absolute and relative. Absolute bodhicitta is the direct insight into the nature of mind. Within absolute bodhicitta, or the absolutely awakened mind, there is no distinction between subject and object, self and other; all sentient beings are spontaneously recognized as perfect manifestations of Buddha nature. Very few people are capable of experiencing absolute bodhicitta right away, however. I certainly wasn't. Like most people, I needed to train along the more gradual path of relative bodhicitta.

There are several reasons why this path is referred to as "relative." First, it is related to absolute bodhicitta in the sense that it shares the same goal: the direct experience of Buddha nature, or awakened mind. To use an analogy, absolute bodhicitta is like the top floor of a building, while relative bodhicitta may be compared to the lower floors. All the floors are part of the same building, but each of the lower floors stands in a relative relationship to the top floor. If we want to reach the top floor, we have to pass through all of the lower floors. Second, when we've achieved the state of absolute bodhicitta, there is no distinction between sentient beings; every living creature is understood as a perfect manifestation of Buddha nature. In the practice of relative bodhicitta, however, we're still working within the framework of a relationship between subject and object or self and other. Finally, according to many great teachers, such as Jamgön Kongtrul in his book *The Torch of Certainty,* development of absolute bodhicitta depends on developing relative bodhicitta.[1]

Developing relative bodhicitta always involves two aspects: aspiration and application. *Aspiration bodhicitta* involves cultivating the heartfelt desire to raise all sentient beings to the level at which they recognize their Buddha nature. We begin by thinking, *I wish to attain complete awakening in order to help all sentient beings attain the same state.* Aspiration bodhicitta focuses on the fruit, or the result, of practice. In this sense, aspiration bodhicitta is like focusing on the goal of carrying everyone to a certain destination—for example, London, Paris, or Washington, D.C. In the case of aspiration bodhicitta, of course, the "destination" is the total awakening of the mind, or absolute bodhicitta. *Application bodhicitta*—often compared in classic texts to actually taking the steps to arrive at an intended destination— focuses on the path of attaining the goal of aspiration bodhicitta: the liberation of all sentient beings from all forms and causes of suffering through recognition of their Buddha nature.

As mentioned, while practicing relative bodhicitta, we're still caught up in regarding other sentient beings from a slightly dualistic perspective, as if their existence were relative to our own. But when we generate the motivation to lift not only ourselves but all sentient beings to the level of complete recognition of Buddha nature, an odd thing happens: The dualistic perspective of "self" and "other" begins very gradually to dissolve, and we grow in wisdom and power to help others as well as ourselves.

As an approach to life, cultivating relative bodhicitta is certainly an improvement on the way we ordinarily deal with others, though it does take a certain amount of work. It's so easy to condemn other people who don't agree with our own point of view, isn't it? Most of us do so as easily and unthinkingly as smashing a mosquito, a cockroach, or a fly.

The essence of developing relative bodhicitta is to recognize that the desire to squash a bug and the urge to condemn a person who disagrees with us are fundamentally the same. It's a fight-or-flight response deeply embedded in the reptilian layer of our brains—or, to put it more bluntly, our crocodile nature.

So the first step in developing relative bodhicitta is to decide, "Would I rather be a crocodile or a human being?"

Certainly there are advantages to being a crocodile. Crocodiles are very good at outsmarting their enemies and simply surviving. But they cannot love or experience being loved. They don't have friends. They can never experience the joys of raising children. They have very little appreciation for art or music. They can't laugh. And many of them end up as shoes.

If you've gotten this far in reading this book, chances are you're not a crocodile. But you've probably met a few people who act like crocodiles. The first step in developing relative bodhicitta is to let go of your distaste for "crocodilelike" people and cultivate some sense of compassion toward them, because they don't recognize how much of the richness and beauty of life they're missing. Once you can do that, extending relative bodhicitta toward all sentient beings—including real crocodiles and whatever other living creatures might annoy, frighten, or disgust you—becomes a lot easier. If you just take a moment to think about how much these creatures are missing out on, your heart will almost automatically open up to them.

Actually, aspiration bodhicitta and application bodhicitta are like two sides of the same coin. One can't exist without the other. Aspiration bodhicitta is the cultivation of an unrestricted readiness to help all living beings achieve a state of complete happiness and freedom from pain and suffering. Whether you're actually able to free them doesn't matter. The important thing is your intention. Application bodhicitta involves the activities required to carry out your intention. Practicing one aspect strengthens your ability to cultivate the other.

There are many ways to practice application bodhicitta: for example, trying your best to refrain from stealing, lying, gossiping, and speaking or acting in ways that intentionally cause pain; acting generously toward others; patching up quarrels; speaking gently and calmly rather than "flying off the handle"; and rejoicing in the good things that happen to other people rather than allowing yourself to become overwhelmed by jealousy or envy. Conduct of this sort is a means of extending the experience of meditation into every aspect of daily life.

There is no greater inspiration, no greater courage, than the intention to lead all beings to the perfect freedom and complete well-being of recognizing their true nature. Whether you accomplish this intention isn't important. The intention alone has such power that as you work with it, your mind will become stronger; your mental afflictions will diminish; you'll become more skillful in helping other beings; and in so doing, you'll create the causes and conditions for your own well-being.

14

THE HOW, WHEN, AND
WHERE OF PRACTICE

*Pure strong confidence . . . is something to be built up
gradually.*

—The Twelfth Tai Situ Rinpoche,
Tilopa: Some Glimpses of His Life,
translated by Ken Holmes

ONE OF THE questions I'm often asked is "Why are there so many
methods, and which is the right one for me?"

If you look around, you can't help but recognize that no two people
are exactly the same in terms of their temperaments and capabilities.
Some people are very good with words; they easily understand verbal
instructions and are comfortable explaining things verbally to others.
Other people are more "visual"; they understand things best when
they're explained with diagrams and pictures. Some people have a
keener sense of hearing than others, while some have a better sense of
smell. Some people are analytical and find it very easy to work out
complicated mathematical formulas. Others are "poets at heart," extremely adept at explaining the world to themselves and others
through metaphors and analogies.

CHOOSE WHAT WORKS FOR YOU

Post the watchman of mindfulness, and rest.

—GYALWA YANG GÖNPA,
translated by Elizabeth M. Callahan

Different circumstances require different measures, so it's always helpful to have a number of options at your disposal. This principle applies to virtually every aspect of life. For example, in business or personal relationships, it's sometimes better to take the time to compose, revise, and communicate your ideas via e-mail, while at other times a phone call or even a face-to-face meeting would be more effective.

Similarly, in terms of meditation, the most appropriate technique depends as much on the particular situation as on your personal temperament and capabilities. When you're dealing with emotions like sadness, anger, or fear, for example, sometimes tonglen practice might be the best approach. Sometimes simply using the emotion itself as a focus for basic shinay practice might work better. Often the only way to find the technique that works best for you is through trial and error.

The main point is to choose whichever method appeals most to you and work with it for a while. If you're more of a "visual" person, try working with form meditation as you begin to practice calming your mind. If you're the type of person who is more alert to physical sensations, try working on scanning your body or focusing on your breath. If you're a "verbal" type, try working with a mantra. The technique itself doesn't matter. The important thing is to learn how to rest your mind—to work *with* it instead of being worked *by* it.

Because the mind is so active, though, it's easy to get bored with just one method. After a couple of days, weeks, or months of working with a particular practice, it's easy to find yourself thinking, *Oh no, I have to meditate again.* Suppose, for example, you start off meditating on form. At first it seems very nice, very soothing. Then one day, for no reason at all, you're just tired of it. You hate the whole idea of meditating on form. That's fine. You don't have to meditate on form anymore. You can try something else, like meditating on sound.

For a while, the new approach seems very fresh and really exciting. You might find yourself thinking, *Wow, I never felt this clear before!* Then, after a while, you might find yourself growing bored with the new technique you've adopted. That's fine, too. If you become bored with meditating on sound, you can try something new, like meditating on smell, or watching your thoughts, or bringing your attention to your breath.

Now does it become clear why the Buddha taught so many different approaches to meditation? Even before the invention of television, the Internet, radios, MP3 players, and telephones, he *knew* how restless the human mind was—and how desperate it was for distraction. So he gave us a lot of ways to work with the restless nature of the mind.

Whatever method you choose, it's very important during each session to alternate between focusing on an object and simply resting your mind in objectless meditation. The point of working with supports for meditation is to develop a degree of mental stability that allows you to be *aware of your own mind as it perceives things*. Resting your mind between objectless meditation and object-based meditation gives you a chance to assimilate whatever you have experienced. By alternating between these two states, no matter what situation you find yourself in—whether you're dealing with your own thoughts and emotions or with a person or a situation that appears "out there"—you'll gradually learn to recognize that whatever is going on is intimately connected with your own awareness.

SHORT PERIODS, MANY TIMES

> *Be free from all striving.*
>
> —Tilopa, *Ganges Mahāmudrā*,
> translated by Elizabeth M. Callahan

Establishing a formal practice is one of the most effective ways to cut through the long-established neuronal gossip that creates the perception of an independent or inherently existing "self" and independently or inherently existing "others." When you set aside time for formal practice, you develop a constructive habit that not only weakens old neuronal patterns, but effectively succeeds in establishing new patterns that enable you to recognize the participation of your own mind in how you perceive.

Though you can practice formally at any time of day, I was taught that the best period to begin formal practice is first thing in the morning after a good night's sleep, at which point the mind is most refreshed and relaxed, before getting involved with all the daily stuff. Taking the time to practice before you leave the house for work or to run whatever errands you have to do sets the tone for your entire day, and also reinforces your own commitment to practice throughout the day.

For some people, though, meditating formally at the beginning of the day simply isn't possible, and trying to force a period of early-morning meditation into your schedule will tend to make you think of meditation as a chore. If you find that to be the case, by all means choose a more convenient time—perhaps at lunchtime, after dinner, or just before going to bed.

There are no "rules" governing formal practice. But there is one very practical guideline, which my father emphasized again and again to all of his students in a way that would make it easy for us to remember: *Short periods, many times.*

When I began teaching, I found that many students who were just starting out in meditation tended to set unrealistic goals for themselves. They felt that in order to meditate they had to sit in perfect meditation posture for as long as humanly possible. So they sat there, trying to "lock in" to meditation, trying to will themselves into a state of serenity. For a few seconds this approach seemed to work: They actually did feel some calmness. But the mind is always moving, always processing new ideas, new perceptions, and new sensations. That's its job. Meditation is about learning to work *with* the mind *as it is,* not about trying to force it into some sort of Buddhist straitjacket.

We think we're being diligent by sitting down to meditate for hours at a stretch. But real diligence doesn't mean forcing yourself beyond your natural limits; it means simply trying to do your best, rather than focusing on the result of what you're trying to accomplish. It means finding a comfortable middle ground between being too relaxed and too wound up.

The sutras record another story about an accomplished sitar player who was a direct student of the Buddha. The Buddha found this particular man especially difficult to teach because his mind was either too tight or too loose. When he was too tight, he would not only be unable to meditate, but also could never remember how to recite the simple prayers the Buddha had taught him. When he allowed his mind to become too loose, he would simply stop practicing altogether and fall asleep.

So the Buddha finally asked him, "What do you do when you go to your house? Do you play your musical instrument?"

The student replied, "Yes, I do play."

"Are you good at it?" the Buddha asked.

The player answered, "Yes, I am actually the best in this country."

"So how do you play?" the Buddha inquired. "When you are playing this music, how do you tune your instrument? Do you make the strings very tight or very loose?"

"No," the student replied. "If I tighten the strings too much, they make a sound like *tink, tink, tink.* If I don't tighten them enough, they make a sound like *blump, blump, blump.* The chord is properly tuned when I reach a point that is in balance, neither too tight nor too loose."

The Buddha smiled and exchanged a long look with the sitar player. Finally he said, "That's exactly what you have to do with meditation." This story illustrates the importance of avoiding undue tension when you first begin meditation practice. Given the busy schedules most people have nowadays, setting aside even fifteen minutes a day at the beginning for formal practice represents a substantial commitment. Whether you divide it up into three five-minute sessions or five three-minute sessions doesn't matter.

Especially in the beginning, it's absolutely essential to spend whatever time you can in practice, without undue strain. The best advice I can offer is to approach meditation practice the way some people approach going to the gym. It's far better to spend fifteen minutes working out at the gym than not working out at all. Even the fifteen minutes you can dedicate to practice is better than spending no time at all. Some people can lift only ten pounds, while others can easily lift fifty. Don't lift fifty if you can lift only ten; otherwise you'll strain yourself and probably stop. And just as with working out at the gym, when you meditate, do the best you can. Don't go beyond your personal limits. *Meditation is not a competition. The fifteen minutes you spend lightly in meditation practice may in the end prove much more beneficial than the hours spent by people trying too hard by practicing for longer periods of time.* In fact, the best rule is to spend *less* time meditating than you think you can. If you think you can practice for four minutes, stop at three; if you think you can practice for five minutes, stop at four. Practicing in this way, you'll find yourself eager to begin again. Rather than thinking you've accomplished your goal, leave yourself wanting more.

Another way to make your short periods of formal practice go even faster is to spend a few moments generating bodhicitta, the desire to attain some degree of realization for the benefit of others. Don't worry about whether the desire is especially strong; the motivation alone is sufficient, and after working at it for a while, you'll probably begin to find that the desire has taken on a real significance, a deeply personal meaning.

After spending a few moments generating this openhearted attitude, allow your mind to rest for a few moments in objectless meditation. This is important no matter which method you choose to work with for a particular session.

Between resting your mind and generating bodhicitta, at least a minute has already passed. Now you have a good minute and a half to work with whatever practice you've chosen, whether it's focusing on a visual object, a smell, or a sound; looking at your thoughts or feelings; or practicing some form of compassion meditation. Then just rest your mind in objectless shinay for half a minute or so.

And at the end of your practice, you've got about thirty seconds or so to do what in many Western languages has been translated as "dedicating the merit." A question that comes up a lot, both in public teachings and in private interviews with my personal students, is "Why should we bother taking this final step of dedicating merit?"

Dedicating merit at the end of any practice is an aspiration that whatever psychological or emotional strength you've gained through practice be passed on to others—which is not only a wonderful short compassion practice but also an extremely subtle way of dissolving the distinction between "self" and "others." Dedication of merit takes about thirty seconds, whether you recite it in Tibetan or English. In English, a rough translation goes like this:

> By this power, may all beings,
> Having accumulated strength and wisdom,
> Achieve the two clear states
> That arise from strength and wisdom.

Some schools of thought—for which, I admit, there is no scientific proof—hold that since the actual sound waves of the prayer as recited in Tibetan have been reverberating through the centuries, saying the prayer in the original language may enhance its power by connecting the recitation with those ancient reverberations. With that in mind, I offer you a rough transliteration:

> *Gewa di yee che wo kun*
> *Sönam ye shay tsok dsok nay*
> *Sönam ye shay lay jung wa*
> *Tampa ku nyee top par shok.*

Whether you choose to end your practice using the formal Tibetan or the less-formal English, there's a very practical reason for completing your practice by dedicating *sönam*—a Tibetan term that means "mental strength" or "the ability to develop mental strength." When we do something nice, our natural tendency is to think, *What a good person I am! I've just meditated. I've just made an aspiration for all beings everywhere to experience real happiness and to be free from suffering. What am I going to get out of this? How is my life going to change for the better? What's this going to do for* me?

These might not be the exact words passing through your mind, but something similar probably does.

And, really, you have done something good.

The only problem is that congratulating yourself in this way tends to emphasize a sense of difference between yourself and others.

Thoughts like "*I've* done something good," "What a good person *I* am," or "*My* life is going to change" subtly reinforce the idea of yourself as separate from other beings—which, in turn, undermines whatever sense of compassion, confidence, and safety your practice may have generated.

By dedicating the merit of your practice—in other words, by deliberately generating the thought that, consciously or unconsciously, *everyone* shares the desire for peace and contentment among all sentient beings—you very subtly dissolve the neuronal habit of perceiving any sort of difference between ourselves and others.

INFORMAL PRACTICE

> *In the middle of working, remember to recognize the essence of the mind.*
>
> —TULKU URGYEN RINPOCHE, *As It Is*, Volume 1,
> translated by Erik Pema Kunsang

Sometimes it's just impossible to make time for formal practice every day. You might have to spend hours preparing for a crucial business meeting, or perhaps you might have to attend an important event, like a wedding or a birthday party. Sometimes you've promised to do something special with your children, your partner, or your spouse. Sometimes you're just so tired from everything you had to do during the week you just want to spend the day in bed or watching TV.

Will skipping a day or two of formal practice make you a bad person? No. Will it reverse whatever changes you've made when you had the time to devote to formal practice? No. Will skipping a day or two (or three) of formal practice mean you have to start all over again working with an untamed mind? No.

Formal practice is great, because sitting for five, ten, or fifteen minutes a day creates an opportunity to begin changing your perspective. But most of the Buddha's early students were farmers, shepherds, and nomads. Between taking care of their crops or animals and looking after their families, they didn't have a lot of time to sit down nicely with their legs crossed, their arms straight, and their eyes correctly focused for even five minutes of formal practice. There was always a sheep bleating somewhere, or a baby crying, or someone rushing into their tent or hovel to say that a sudden rain was about to ruin their crops.

The Buddha understood these problems. Although the fanciful stories about his birth and upbringing describe him as the son of a wealthy king who raised him in a fabulous pleasure palace, his origins were actually a lot more humble. His father was only one of several chieftains of the sixteen republics battling to resist being swallowed up by the powerful Indian monarchy. His mother died giving birth to him; his father forced him to marry and produce an heir when he was just a teenager. He was disinherited when he ran away from home to pursue a life that might have a deeper meaning than political and military scheming.

So, when we talk about the Buddha, we're talking about a man who understood that life doesn't always afford the opportunity or leisure to practice formally. One of his greatest gifts to humanity was the lesson that it's possible to meditate anytime, anywhere. In fact, bringing meditation into your daily life is one of the main objectives of Buddhist practice. Any daily activity can be used as an opportunity for meditation. You can watch your thoughts as you go through your day, rest your attention momentarily on experiences like taste, smell, form, or sound, or simply rest for a few seconds on the marvelous experience of simply being aware of the experiences going on in your mind.

When practicing informally, though, it's important to set some sort of goal for yourself—for example, twenty-five informal meditation sessions lasting no more than a minute or two throughout the day. It's also helpful to keep track of your sessions. Monks and nomads in the Third World often keep track using prayer beads. But people in the West have a much wider range of options—including portable calculators, PDAs, and even those little counting machines people use in grocery stores. You can also keep track of your sessions simply by writing them down on a notepad. The main thing is to count every informal meditation practice so you can track it against your goal. For example, if you're using objectless meditation, count it as one. Then you lose it, try again, and count it as two.

One of the great benefits of organizing your meditation practice in this way is that it's convenient and portable. You can practice anywhere—on the beach, at the movies, on the job, in a restaurant, on the bus or subway, or at school—as long as you remember that your intention to meditate *is* meditation. No matter what your opinion might be about how well you've meditated, the point is to keep track of your intentions to meditate. When you come up against resistance, just remember the story about how the old cow pees while walking along throughout the day. That should be enough to bring a smile to your face and remind you that practicing is as easy, and as necessary, as relieving yourself.

Once you're comfortable with twenty-five short sessions a day, you can push your goal up to fifty informal sessions, then gradually up to a hundred. The main thing is to make a plan. If you don't, you'll forget about practice altogether. Those few seconds or minutes out of each day during which you allow yourself to rest or focus help you to stabilize your mind, so that when you finally do get a chance to practice formally, it won't be like sitting down to dinner with a stranger. You'll find your thoughts, feelings, and perceptions much more familiar, like old friends you can sit down with and talk to honestly.

There are a couple of other benefits to informal practice. First, when you integrate practice into your daily life, you avoid the trap of being calm and peaceful during formal meditation and then turning around and being tense and angry at the office. Second, and perhaps more important, practicing informally in daily life gradually eradicates the all-too-common misconception that you have to be someplace absolutely quiet in order to meditate.

No one in history has ever found such a place. Distractions are everywhere. Even if you climb to the top of a mountain, at first you may feel some relief in the relative silence there compared to the sounds of the city or the office. But as your mind settles, you'll undoubtedly begin to hear small sounds, like crickets chirping, the wind rustling through leaves, birds or small animals poking around, or water dripping through rocks—and suddenly the great silence you were seeking is interrupted. Even if you try meditating indoors, closing all the windows and doors, you're bound to be distracted by something—itching, back pain, the need to swallow, the sound of water dripping from a faucet, a clock ticking, or the noise of someone walking on the floor above. No matter where you go, you'll always find distractions. The greatest benefit of informal practice lies in learning how to deal with these distractions, no matter what form they take and no matter how irritating they are.

ANYTIME, ANYWHERE

Join whatever you meet with meditation.

—Jamgön Kongtrul, *The Great Path of Awakening*,
translated by Ken McLeod

With this in mind, let's take a look at some of the ways you can practice during daily life, and even use what might ordinarily seem to be distractions as supports for resting your mind. The old texts call this "taking your life as the path."

Simply walking down the street can be a great opportunity to develop mindfulness. How often do you find yourself setting out on a particular task, like heading out to the grocery store or walking to a restaurant for lunch, and find yourself at your destination without even realizing how you got there? This is a classic example of allowing the crazy monkey to run riot, spinning out all sorts of distractions that not only keep you from experiencing the fullness of the present moment, but also rob you of the chance to focus and train your awareness. The opportunity here is to decide consciously to bring your attention to your surroundings. Look at the buildings you pass, at other people on the sidewalk, at the traffic in the streets, at the trees that may be planted along your route. When you pay attention to what you see, the crazy monkey settles down. Your mind becomes less agitated, and you begin to develop a sense of calmness.

You can also bring your attention to the physical sensation of walking, to the feeling of your legs moving, your feet touching the ground, the rhythm of your breathing or your heartbeat. This works even if you're in a rush, and is actually a great method of combating the anxiety that generally accompanies trying to get somewhere in a hurry. You can still walk quickly while bringing your attention to either your physical sensations or the people, places, or things you pass along your way. Just allow yourself to think, *Now I'm walking down the street. . . . Now I'm seeing a building. . . . Now I'm seeing a person in a T-shirt and jeans. . . . Now my left foot is touching the ground. . . . Now my right foot is touching the ground. . . .*

When you bring conscious awareness to your activity, distractions and anxieties will gradually fade and your mind will become more peaceful and relaxed. And when you do arrive at your destination, you'll be in a much more comfortable and open position to deal with the next stage of your journey.

You can bring the same sort of attention to driving or to day-to-day experiences in your home or your workplace, simply by bringing your attention to various objects in your visual field, or using sounds as supports. Even simple tasks like cooking and eating provide opportunities for practice. While chopping vegetables, for example, you can bring your attention to the shape or color of each piece as you chop it, or to the sounds of soup or sauce bubbling. While eating, bring your attention to the smells and tastes you experience. Alternatively, you can practice objectless meditation in any of these situations, allowing your mind to rest simply and openly as you go about any activity, without attachment or aversion.

You can even meditate while sleeping or dreaming. As you fall asleep, you can either rest your mind in objectless meditation or gently rest your attention on the feeling of sleepiness. Alternatively, you can create an opportunity to turn your dreams into meditation experiences by reciting silently to yourself several times as you fall asleep, *I will recognize my dreams, I will recognize my dreams, I will recognize my dreams.*

IN CONCLUSION

> *When you begin to feel completely desolate, you begin to help yourself, you make yourself at home.*
>
> —CHÖGYAM TRUNGPA, *Illusion's Game*

Meditation isn't a one-size-fits-all practice. Every individual represents a unique combination of temperament, background, and abilities. Recognizing this, the Buddha taught a variety of methods to help people in all walks of life and in any situation recognize the nature of their minds and true freedom from the mental poisons of ignorance, attachment, and aversion. Mundane as some of these methods may appear, they actually represent the heart of Buddhist practice.

The essence of the Buddha's teachings was that while formal practice can help us to develop direct experience of emptiness, wisdom, and compassion, such experiences are meaningless unless we can bring them to bear on every aspect of our daily lives. For it's in facing the challenges of daily life that we can really measure our development of calmness, insight, and compassion.

Even so, the Buddha invited us to try the practices for ourselves. In one of the sutras, he urged his students to test his teachings through practice, rather than accepting them simply at face value:

> As you would burn, cut, and rub gold,
> Likewise, the wise monk examines my teachings.
> Examine my teachings well,
> But don't take them on faith.

In the same spirit, I ask you to try the teachings for yourselves to see if they work for you. Some of the practices may help you; some may not. Some of you may find a certain affinity with one or more techniques right away, while other methods require a bit more practice. Some of you may even find that meditation practice does not benefit you at all. That's okay, too. The most important thing is to find and work with a practice that produces a sense of calmness, clarity, confidence, and peace. If you can do that, you will benefit not only yourselves, but also everyone around you; and that is the goal of every scientific or spiritual practice, isn't it? To create a safer, more harmonious, and gentle world, not only for ourselves, but for generations to come.

PART THREE

THE FRUIT

Experience changes the brain.
—JEROME KAGAN, *Three Seductive Ideas*

15

PROBLEMS AND POSSIBILITIES

*At the beginning our mind is not able to remain stable
and rest for very long. However, with perseverance and
consistency, calm and stability gradually develop.*
— BOKAR RINPOCHE, *Meditation: Advice to Beginners,*
translated by Christiane Buchet

WONDERFUL EXPERIENCES CAN occur when you rest your mind in
meditation. Sometimes it takes a while for these experiences to occur;
sometimes they happen the very first time you sit down to practice.
The most common of these experiences are bliss, clarity, and nonconceptuality.

Bliss, in the way it was explained to me, is a feeling of undiluted
happiness, comfort, and lightness in both the mind and the body. As
this experience grows stronger, it seems as if everything you see is
made of love. Even experiences of physical pain become very light and
hardly noticeable at all.

Clarity is a sense of being able to see into the nature of things as
though all reality were a landscape lit up on a brilliantly sunny day
without clouds. Everything appears distinct and everything makes
sense. Even disturbing thoughts and emotions have their place in this
brilliant landscape.

Nonconceptuality is an experience of the total openness of your
mind. Your awareness is direct and unclouded by conceptual distinctions such as "I" or "other," subjects and objects, or any other form of
limitation. It's an experience of pure consciousness as infinite as

space, without beginning, middle, or end. It's like becoming awake within a dream and recognizing that everything experienced in the dream isn't separate from the mind of the dreamer.

Very often, though, what I hear from people who are just starting out to meditate is that they sit there and nothing happens. Sometimes they feel a very brief, very slight sense of calmness. But in most cases they don't feel any different from the way they did before they sat down or after they got up. That can be a real disappointment.

Some people, furthermore, feel a sense of disorientation, as though their familiar world of thoughts, emotions, and sensations has tilted slightly—which may be pleasant or unpleasant.

As I've mentioned before, whether you experience bliss, clarity, disorientation, or nothing at all, *the intention to meditate is more important than what happens when you meditate.* Since mindfulness is already present, just making the effort to connect with it will develop your awareness of it. If you keep practicing, gradually you might feel a little something, a sense of calmness or peace of mind that is slightly different from your ordinary state of mind. When you begin to experience that, you'll intuitively understand the difference between the distracted mind and the undistracted mind of meditation.

In the beginning, most of us aren't able to rest our minds in bare awareness for a very long time at all. If you can rest for only a very short time, that's fine. Just follow the instruction given earlier to repeat that short period of relaxation many times over in any given session. Even resting the mind for the time it takes you to breathe in and breathe out is enormously useful. Just do that again and again and again.

Conditions are always changing, and real peace lies in the ability to adapt to the changes. For example, suppose you're sitting there focusing calmly on your breathing and your upstairs neighbor starts vacuuming, or a dog starts barking somewhere in the vicinity. Maybe your back or your legs start to hurt, or maybe you feel an itch. Or maybe the memory of a fight you had the other day pops into your head for no apparent reason. These things happen all the time—and that's another reason why the Buddha taught so many different methods of meditation.

When distractions of this sort occur, just make them a part of your practice. Join your awareness to the distraction. If your breathing practice is interrupted by the noise of a dog barking or a vacuum cleaner, switch to sound meditation, resting your attention on the noise. If you feel pain in your back or legs, bring your attention to the mind that feels the pain. If you have an itch, go ahead and scratch it. If you ever have a chance to sit in a Buddhist shrine room while a lecture or a chanting practice is taking place, you'll undoubtedly see the monks restlessly scratching themselves, shifting on their cushions, or coughing. But, chances are, if they've taken their training seriously enough, they're shifting around, scratching, and so on mindfully—bringing their attention to the sensation of the itch, the sensation of scratching it, and the relief they feel when they're done scratching.

If you're distracted by strong emotions, you can try focusing, as was taught earlier, on the mind that experiences the emotion. Or you might try switching to tonglen practice, using whatever you're feeling—anger, sadness, jealousy, desire—as the basis for the practice.

At the opposite end of the scale, a number of people I know find their minds getting foggy or sleepy when they practice. It's such an effort just to keep their eyes open and their attention focused on what they're doing. The thought of giving up for the day and flopping down in bed seems very tempting.

There are a couple of ways to deal with this situation. One, which is simply a variation on being mindful of physical sensations, is to rest your attention on the sensation of dullness or sleepiness itself. In other words, use your dullness rather than being used by it. If you can't sit up, just lie down while keeping your spine as straight as possible.

Another remedy is simply to raise your eyes so that you're gazing upward. You don't have to lift your head or your chin, just turn your gaze upward. That often has the effect of waking your mind. Lowering your gaze, meanwhile, can have a calming effect when your mind is agitated.

If none of the remedies for dullness or distraction work, I usually recommend to students that they just stop for a while and take a break. Go for a walk, take care of something around the house, exercise, read a book, or work in the garden. There's no sense in trying to make yourself meditate if your mind and body aren't willing to cooperate. If you try to hammer away at your resistance, eventually you'll get frustrated with the whole idea of meditation and decide to chuck it in favor of achieving happiness through some temporary attraction. At those times, all those channels available through your satellite dish or cable TV box look pretty promising.

PROGRESSIVE STAGES OF MEDITATION PRACTICE

> *Allow the water [of mind] clouded by thoughts to clear.*
> —Tilopa, *Ganges Mahāmudrā,*
> translated by Elizabeth M. Callahan

When I first began meditating, I was horrified to find myself experiencing *more* thoughts, feelings, and sensations than I had before I began practicing. It seemed that my mind was becoming more agitated rather than more peaceful. "Don't worry," my teachers told me. "Your mind isn't getting worse. Actually, what's happening is that you're simply becoming more aware of activity that has been going on all the time without your noticing it."

The Waterfall Experience

They explained this experience through the analogy of a waterfall suddenly swollen by a spring thaw. As melting snow floods down from the mountains, they told me, all sorts of stuff gets stirred up. There might be hundreds of rocks, stones, and other elements flowing through the water, but it's impossible to see them all because the water is rushing by so quickly, shaking up all sorts of debris that clouds the water; and it's very easy to become distracted by all that mental and emotional debris.

   They taught me a short prayer known as the *Dorje Chang Tungma,* which I found very helpful when my mind seemed overwhelmed by thoughts, emotions, and sensations. Part of it, roughly translated, goes like this:

> The body of meditation, it's said, is nondistraction.
> Whatever thoughts are perceived by the mind are nothing in themselves.
> Help this meditator who rests naturally in the essence
> of whatever thoughts arise to rest in the mind as it naturally is.

In working with many students around the world, I've observed that the "waterfall" experience is often the first one people encounter when they begin to meditate. There are actually several common reactions to this experience, and I've experienced them all. In a sense, I consider myself fortunate, because having experienced these stages has enabled me to develop greater empathy toward my students. At the time, though, the waterfall seemed like a terrible ordeal.

The first variation involves trying to stop the waterfall by deliberately trying to block thoughts, feelings, and sensations in order to experience a sense of calmness, openness, and peace. This attempt to block experience is counterproductive, because it creates a sense of mental or emotional tightness that ultimately manifests as a physical strain, especially in the upper body: Your eyes roll upward, your ears become taut, and your neck and shoulders become abnormally tense. I tend to think of this phase of practice as "rainbowlike" meditation, because the calmness after blocking the flow of the waterfall is as illusory and transient as a rainbow.

Once you let go of trying to impose an artificial sense of calmness, you'll find yourself confronting the "raw" waterfall experience, in which your mind is carried away by various thoughts, feelings, and sensations you may previously have tried to block. This is generally the kind of "Oops" experience described earlier in Part Two—in which you start trying to observe your thoughts, feelings, and sensations, and then get carried away by them. You recognize you've been carried away, and you try to force yourself back to simply observing what's going on in your mind. I call this the "hook" form of meditation, in which you try to hook your experiences, and feel some regret if you allow yourself to be carried away by them instead.

There are two ways to deal with the "hook" situation. If your regret over letting yourself be carried away by distractions is really strong, then just let your mind rest gently in the experience of regret. Otherwise, let go of the distractions and rest your awareness in your present experience. You might, for example, try bringing attention to your physical sensations: Perhaps your head is a little bit warm, your heart is beating a little faster, or your neck or shoulders are a little tense. Just rest your awareness on these or other experiences that occur in the present moment. You might also try simply resting with bare attention—as discussed in Parts One and Two—in the rush of the waterfall itself.

However you deal with it, the waterfall experience provides an important lesson in letting go of preconceived ideas about meditation. The expectations you bring to meditation practice are often the greatest obstacles you will encounter. The important point is to allow yourself simply to be aware of whatever is going on in your mind as it is.

Another possibility is that experiences come and go too quickly for you to recognize them. It's as though each thought, feeling, or sensation is a drop of water that falls into a large pool and is immediately absorbed. That's actually a very good experience. It's a kind of objectless meditation, the best form of calm-abiding practice. So if you can't catch every "drop," don't blame yourself—congratulate yourself, because you've spontaneously entered a state of meditation that most people find hard to reach.

After a little bit of practice, you'll find that the rush of thoughts, emotions, and so on begins to slow, and it becomes possible to distinguish your experiences more clearly. They were there all the time, but as in the case of a real waterfall, in which the rush of water stirs up so much dirt and sediment, you just couldn't see them. Likewise, as the habitual tendencies and distractions that normally cloud the mind begin to settle through meditation, you'll begin to see the activity that has been going on all the time just below the level of ordinary awareness.

Still, you might not be able to observe each thought, feeling, or perception as it passes, but only catch a fleeting glimpse of it—rather like the experience of having just missed a bus, as described earlier. That's okay, too. That sensation of just having missed observing a thought or feeling is a sign of progress, an indication that your mind is sharpening itself to catch traces of movement, the way a detective begins to notice clues.

As you keep practicing, you'll find that you're able to become aware of each experience more clearly as it occurs. The analogy my teachers provided to describe this phenomenon was that of a flag waving in a strong wind. The flag shifts and moves constantly, according to the direction of the wind. The movements of the flag are like the events whipping through your mind, while the flagpole is like your natural awareness: straight and steady, never shifting, anchored to the ground. It doesn't move, no matter how strong the wind that whips the flag in one direction or another.

The River Experience

Gradually, as you continue to practice, you'll inevitably find yourself able to clearly distinguish the movements of thoughts, emotions, and sensations through your mind. At this point you've begun to shift from the "waterfall" experience to what my teachers called the "river" experience, in which things are still moving, but more slowly and gently. One of the first signs that you've entered the river phase of meditation experience is that you'll find yourself occasionally entering a state of meditative awareness without much effort, naturally joining your awareness with whatever is going on inside or around you. And when you sit in formal practice, you'll have clearer experiences of bliss, clarity, and nonconceptuality.

Sometimes the three experiences occur simultaneously, and sometimes one experience is stronger than the other two. You may feel that your body is becoming lighter and less tense. You may find that your perceptions are becoming clearer and, in a way, more "transparent," in that they don't seem so heavy or oppressive as they may have felt in the past. Thoughts and feelings don't seem so powerful anymore; they become infused with the "juice" of meditative awareness, appearing more like passing impressions than absolute facts. When you "enter the river," you'll find your mind becoming calmer. You'll find yourself not taking its movements so seriously, and as a result you'll find yourself spontaneously experiencing a greater sense of confidence and openness, which won't be shaken by who you meet, what you experience, or where you go. Even though such experiences may come and go, you will begin to sense the beauty of the world around you.

Once that starts to happen, you'll also begin to discern tiny gaps between experiences. At first the gaps will be very short—just fleeting glimpses of nonconceptuality or nonexperience. But over time, as your mind becomes calmer, the gaps will become longer and longer. This is really the heart of shinay practice: the ability to notice and rest in the gaps between thoughts, emotions, and other mental events.

The Lake Experience

During the "river" experience, your mind may still have its ups and downs. When you reach the next stage, which my teachers called the "lake" experience, your mind begins to feel very smooth, wide, and open, like a lake without waves. You find yourself genuinely happy, without any ups and downs. You're full of confidence, stable, and you experience a more or less continuous state of meditative awareness, even while sleeping. You may still experience problems in your life—negative thoughts, strong emotions, and so on—but instead of being obstacles, they become further opportunities to deepen your meditative awareness, the way a runner uses the challenge of going half a mile farther to break through a "wall" of resistance and achieve even greater strength and ability.

At the same time, your body begins to feel the lightness of bliss, and your clarity improves so that all your perceptions begin to take on a sharper, almost transparent quality, like reflections in a mirror. While the crazy monkey mind may still pose a few problems during the river phase of experience, when you reach the lake stage, the crazy monkey has retired.

A traditional Buddhist analogy for progressing through these three stages is a lotus rising from the mud. A lotus begins to grow from the mud and sediment at the bottom of a lake or pond, but by the time the flower blooms at the surface, it bears no trace of mud; in fact, the petals actually appear to repel filth. In the same way, when your mind blossoms into the lake experience, you have no trace of clinging or grasping, none of the problems associated with samsara. You might even develop, as did the great masters of old, heightened powers of perception, such as clairvoyance or mental telepathy. If you do have these experiences, though, it's best not to boast about them or mention them to anyone but your teacher or very close students of your teacher.

In the Buddhist tradition, people don't talk much about their own experiences and realizations, mostly because such boasting tends to increase one's own sense of pride and can lead to misuse of the experiences to gain worldly power or influence over other people, which is harmful to oneself and to others. For this reason, training in meditation involves a vow or a commitment—known in Sanskrit as samaya—not to misuse the abilities gained through meditation practice: a vow similar to treaties not to misuse nuclear arms. The consequence of breaking this commitment is the loss of whatever realizations and abilities one has attained through practice.

MISTAKING EXPERIENCE FOR REALIZATION

Give up whatever you're attached to.

—The Ninth Gyalwang Karmapa,
Mahāmudrā: The Ocean of Definitive Meaning,
translated by Elizabeth M. Callahan

Although the lake experience may be considered the crown of shinay practice, it is not in itself realization or full enlightenment. It's an important step along the way, but not the final one. Realization is the full recognition of your Buddha nature, the basis of samsara and nirvana: free from thoughts, emotions, and the phenomenal experiences of sense consciousness and mental consciousness; free from dualistic experiences of self and other, subject and object; infinite in scope, wisdom, compassion, and ability.

My father once told a story about a time when he was still living in Tibet. One of his students, a monk, went up to a mountain cave to practice. One day the monk sent my father an urgent message to come visit him. When my father arrived, the monk excitedly told him, "I've become totally enlightened. I can fly. I know it. But, since you're my teacher, I need your permission."

My father recognized that the monk had merely had a glimpse—an experience—of his true nature, and told him quite bluntly, "Forget about it. You can't fly."

"No, no," the monk excitedly replied. "If I jump from the top of the cave—"

"No," my father interrupted.

They argued back and forth like that for a long while, until the monk finally broke down and said, "Well, if you say so, I won't try."

Since it was approaching noon, the monk offered my father lunch. After serving my father, the monk left the cave, and quite soon afterward, my father heard a strange noise—a kind of *BLUMP*—and then, from far below the cave, came a wail: "Please help me! I've broken my leg!"

My father climbed down to where the monk lay, and said, "You told me you were enlightened. Where is your experience now?"

"Forget about my experience!" the monk cried. "I'm in pain!"

Compassionate as always, my father carried the monk to his cave, splinted his leg, and gave him some Tibetan medicine to help heal his injury. But it was a lesson the monk never forgot.

Like my father, my other teachers were always careful to point out the distinction between momentary experience and true realization. Experience is always changing, like the movement of clouds against the sky. Realization—the stable awareness of the true nature of your mind—is like the sky itself, an unchanging background against which shifting experiences occur.

In order to achieve realization, the important thing is to allow your practice to evolve gradually, beginning with very short periods, several times a day. The incremental experiences of calmness, serenity, and clarity you experience during these short periods will inspire you, very naturally, to extend your practice for longer periods. Don't force yourself to meditate when you're too tired or too distracted. Don't avoid practice when the small, still voice inside your mind tells you it's time to focus.

It's also important to let go of any sensations of bliss, clarity, or non-conceptuality you may experience. Bliss, clarity, and nonconceptuality are all very nice experiences, and are clearly signs of having made a profound connection with the true nature of your mind. But there is a temptation, when such experiences occur, to hold on to them tightly and make them last. It's okay to remember these experiences and appreciate them, but if you try to hang on to them or repeat them, you'll eventually end up feeling disappointed and frustrated. I know, because I've felt the same temptation myself, and I've experienced the frustration when I gave in to the temptation. Each flash of bliss, clarity, or nonconceptuality is a spontaneous experience of the mind *as it is* at that particular moment.

When you try to hold on to an experience like bliss or clarity, the experience loses its living, spontaneous quality; it becomes a concept, a dead experience. No matter how hard you try to make it last, it gradually fades. If you try to reproduce it later, you may get a taste of what you felt, but it will only be a memory, not the direct experience itself.

The most important lesson I learned was to avoid becoming attached to my positive experience if it was peaceful. As with every mental experience, bliss, clarity, and nonconceptuality spontaneously come and go. You didn't create them, you didn't cause them, and you can't control them. They are simply natural qualities of your mind. I was taught that when such very positive experiences occur to stop right there, before the sensations dissipate. Contrary to my expectations, when I stopped practicing as soon as bliss, clarity, or some other wonderful experience occurred, the effects actually lasted much longer than when I tried to hang on to them. I also found that I was much more eager to meditate the next time I was supposed to practice.

Even more important, I discovered that ending my meditation practice at the point at which I experienced something of bliss, clarity, or nonconceptuality was a great exercise in learning to let go of the habit of dzinpa, or grasping. Grasping or clinging too tightly to a wonderful experience is the one real danger of meditation, because it's so easy to think that this wonderful experience is a sign of realization. But in most cases it's just a passing phase, a glimpse of the true nature of the mind, as easily obscured as when clouds obscure the sun. Once that brief moment of pure awareness has passed, you have to deal with the ordinary conditions of dullness, distraction, or agitation that confront the mind. And you gain greater strength and progress through working with these conditions than by trying to cling to experiences of bliss, clarity, or nonconceptuality.

Let your own experience serve as your guide and inspiration. Let yourself enjoy the view as you travel along the path. The view is your own mind, and because your mind is already enlightened, if you take the opportunity to rest awhile along the journey, eventually you'll realize that the place you want to reach is the place you already are.

16

AN INSIDE JOB

*Enlightenment is possible only in that one way—from
the inside.*

— THE TWELFTH TAI SITU RINPOCHE, "A Commentary on
the Aspiration Prayer of Mahamudra, the Definitive
Meaning," in *Shenpen Ösel* 2, no. 1 (March 1998)

ONE OF THE great things about teaching around the world is the opportunity to pick up bits and pieces of different languages. There's a particular American expression I like very much, which refers to a type of crime committed by people within a company: an "inside job." The individuals involved in this sort of crime usually feel safe because they think they know about all the crime-prevention precautions put in place by whatever company they're working for. But it often turns out that they don't know everything, and their own actions give them away.

In a way, letting ourselves be controlled by our mental afflictions is an "inside job." The pain we feel when we lose something we're attached to, or when we confront something we'd rather avoid, is a direct result of not knowing everything we could or should know about our own mind. We're caught by our own ignorance, and trying to free ourselves through some sort of external means—which are simply reflections of the dualistic ignorance that got us into trouble in the first place—only makes our prisons close around us more tightly and securely.

Everything I've learned about the biological processes of thought and perception indicates that the only way to break free from the prison of pain is by performing the same type of activity that imprisoned us in the first place. As long as we don't recognize the peace that exists naturally within our own minds, we can never find lasting satisfaction in external objects or activities.

In other words, happiness and unhappiness are "inside jobs."

TO SURVIVE OR TO THRIVE: THAT IS THE QUESTION

From virtue, all happy states arise.

—GAMPOPA, *The Instructions of Gampopa,*
translated by Lama Yeshe Gyamtso

As a child, I was taught that there are two kinds of happiness: temporary and permanent. Temporary happiness is like aspirin for the mind, providing a few hours of relief from emotional pain. Permanent happiness comes from treating the underlying causes of suffering. The difference between temporary and permanent happiness is similar in many ways to the distinction, discussed in Part One, between emotional states and emotional traits. Genetically, it appears that human beings are programmed to seek temporary states of happiness rather than lasting traits. Eating, drinking, making love, and other activities release hormones that produce physical and psychological sensations of well-being. By releasing these hormones, survival-based activities play an important role in ensuring that we survive as individuals, and that the genes we carry are passed on to future generations.

As explained to me, however, the pleasure we feel in such activities is transitory by genetic design. If eating, drinking, making love, and so on were able to produce permanent sensations of happiness, we'd do these things once and then sit back and enjoy ourselves while others took over the tasks involved in perpetuating the species. In strictly biological terms, the drive to survive propels us more strongly toward unhappiness than toward happiness.

That's the bad news.

The good news is that a biological quirk in the structure of our brains enables us to override many of our genetic predispositions. Instead of compulsively repeating the same activities in order to reexperience temporary states of happiness, we can actually train ourselves to recognize, accept, and rest in a more lasting experience of peace and contentment. This "quirk" is actually the highly developed neocortex, the area of the brain that deals with reasoning, logic, and conceptualization.

There are, of course, disadvantages to having a big, complex neocortex. A lot of people can get so bogged down in weighing and reweighing the pros and cons of everything from ending a relationship to the right time to go to the grocery store that they never make any decisions at all. But the ability to choose among different options is an incredible advantage, one that far outweighs any disadvantages.

DIRECTING THE BRAIN

The firewood is not itself the fire.

—NĀGARJUNĀ, *The Fundamental Wisdom of the Middle Way,* translated by Ari Goldfield

It's fairly common knowledge these days that the brain is divided into two halves, left and right. Each half is more or less a mirror image of the other, complete with its own amygdala, hippocampus, and a big frontal lobe that handles much of the rational processes of the neocortex. I've heard people talk casually about being "left-brained" or "right-brained," referring to a popular idea that people in whom the left half of the brain is more active tend to be more analytical or intellectual, while people in whom the right half is more active tend to be more creative or artistic. I don't know if that's true or not. What I have learned, though, is that research over the past few years indicates that in human beings and other highly evolved species (like our friend the crazy monkey), the two frontal lobes play different roles in shaping and experiencing emotions.

During the 2001 Mind and Life Institute meeting in Dharamsala, Professor Richard Davidson presented the results of a study in which people tested at the Waisman Laboratory for Brain Imaging and

Behavior, in Madison, Wisconsin, were shown pictures designed to evoke different kinds of emotions. These pictures ranged from images of a mother tenderly holding a baby to images of accident and burn victims. The subjects were tested several times over the course of two months, with a few weeks between each test. The results clearly showed an increase in activity in the subjects' left prefrontal lobes when shown pictures normally associated with such positive emotions as joy, tenderness, and compassion, while activity in the right prefrontal lobe increased when the subjects were shown images that provoked negative emotions like fear, anger, and disgust.[1]

In other words, there is a strong indication that positive emotions such as happiness, compassion, curiosity, and joy are linked to activity in the left prefrontal lobe of the brain, while negative emotions like anger, fear, jealousy, and hatred are generated in the right prefrontal lobe. Identifying this connection represents a major step forward in understanding the biological foundations of happiness and unhappiness, and may in the long term provide a basis for developing a practical science of happiness. More immediately, it offers an important key to understanding the results of studies that Professor Davidson and Professor Antoine Lutz would later begin to conduct involving people who had undergone different levels of training in meditation and subjects who had no experience in meditation at all.

The first of these studies, described to me as a "pilot study"—that is to say, a sort of test project designed to assist scientists in developing clinical research projects that could be carried out with much more specific criteria and controls—was conducted in 2001. The subject of the pilot study was a monk who had trained for more than thirty years under some of the greatest masters of Tibetan Buddhism. It's important to note that the results of this pilot study cannot be considered conclusive. First and foremost, of course, it takes some time to review the results of the study in order to sort out unforeseen technical issues. Second, reviewing the results of a pilot study helps scientists distinguish between information that may be relevant to the study and information that may not. Third, in the case of working with Tibetan monks, there are certain language difficulties that often hamper clear communication between the subjects and the researchers. Finally, as discussed near the end of Part Two, there is a natural, samaya-based reticence on the part of Tibetan practitioners to describe the exact nature of their experience to anyone but a qualified teacher.

The Madison pilot study was aimed at determining whether the techniques of mental discipline that the subject had learned over three decades of training could produce objectively measurable changes in the activity among various areas of his brain. For the purposes of the experiment, the monk was asked to engage in several different kinds of meditation practice. These included resting his mind on a particular object, generating compassion, and objectless shinay (which the monk involved in the pilot study described as "open-presence" meditation, a description of simply resting in the open presence of the mind without focusing on a particular object). He alternated between a neutral state for sixty seconds and a specific meditation practice for sixty seconds.

During the pilot study, the monk's brain was monitored using an fMRI scanner followed by two rounds of EEGs—the first using 128 electrodes and the second a massive array of 256, far more than the usual number of sensors used in hospitals, which only measure electrical or brain-wave activity just below the scalp. The pictures I saw of the EEG experiments were actually very funny. It looked as though hundreds of snakes had been attached to the monk's head! But the information gathered by all these snakes, when analyzed by the advanced computer programs developed for the Madison lab, provided a map of activity in regions very deep inside the monk's brain.[2]

Though it would take months for computers to sort through all the complicated data generated by the different brain scans, preliminary examinations of the pilot study indicated shifts between large sets of neuronal circuits in the monk's brain that at least suggest a correspondence between the changes in his brain activity and the meditation techniques he was asked to practice. By contrast, similar scans performed on subjects who'd had no meditation training indicated a somewhat more limited ability to direct the activity of their brains voluntarily while performing a specific mental task.

When speaking about this experiment during a recent trip to England, I was told by several people that a test performed by scientists at University College London using magnetic resonance imaging (MRI) technology had shown that London taxicab drivers—who must undergo a two- to four-year training, known as "the Knowledge," through which they learn to navigate the complicated network of streets in that city—have shown a significant growth in the hippocampus region of the brain, the area associated most typically with spatial memory. In very simple terms, this study begins to confirm that repeated experience can actually change the structure and function of the brain.

The ability to recognize the feelings and sensations of others is a property specific to mammals, which are endowed with the limbic region of the brain.[3] There's no doubt that this capacity can sometimes seem more problematic than it's worth. Wouldn't it be nice to just respond to every situation in simple, black-and-white terms of kill or be killed, eat or be eaten? But what an incredible loss this simple approach to existence would be! The limbic region of our brains affords us the capacity to feel love, and the awareness of being loved. It allows us to experience friendship and to form the basic structures of society that provide us with a greater measure of safety and survival, which help ensure that our children and grandchildren will thrive and grow. The limbic system provides us with the capacity to create and appreciate the subtle emotions evoked by art, poetry, and music. Certainly these capacities are complex and cumbersome; but ask yourself the next time you see an ant or a cockroach scuttling across the floor if you would rather live your life in terms of the simple dimensions of fear or flight, or with the more complex and subtle emotions of love, friendship, desire, and appreciation of beauty.

Two distinct but related functions of the limbic system are involved in the development of loving-kindness and compassion. The first is what neuroscientists have identified as *limbic resonance*[4]—a kind of brain-to-brain capacity to recognize the emotional states of others through facial expression, pheromones, and body or muscular position. It's amazing how quickly the limbic area of the brain can process these subtle signs so that we can not only recognize the emotional states of others, but also adjust our own physical responses accordingly. In most cases, if we haven't trained ourselves to pay bare attention to the shifts and changes in observing our minds, the process of limbic resonance occurs unconsciously. This immediate adjustment is a miraculous demonstration of the brain's agility.

The second function is referred to as *limbic revision,* which in simple terms means the capacity to change or revise the neuronal circuitry of the limbic region, either through direct experience with a person like a lama or a therapist, or through direct interaction with a set of instructions involved, say, with repairing a car or building a swing set.[5] The basic principle behind limbic revision is that the neuronal circuitry in this region of the brain is sufficiently flexible to withstand change. To take a very simple example, suppose you were talking to a friend about someone toward whom you feel a romantic attraction, and as you were discussing this person, your friend said something like "Oh my God, no! That's exactly the type of person you've fallen for before, and look at how much pain that last relationship caused you." It may not be your friend's words that cause you to reconsider going forward with the new relationship, but rather his or her tone of voice and facial expressions, which register on a level of awareness that may not necessarily be conscious.

It would seem that meditation—particularly on compassion—creates new neuronal pathways that increase communication between different areas of the brain, leading to what I've heard some scientists refer to as "whole brain functioning."

From a Buddhist perspective, however, I can say that meditation on compassion fosters a broadening of insight into the nature of experience that stems from unchaining the habitual tendency of mind to distinguish between self and other, subject and object—a unification of the analytical and intuitive aspects of consciousness that is both extremely pleasurable and tremendously liberating.

Through training in loving-kindness and compassion toward others, it's possible to integrate the processes of the limbic region with a more conscious awareness. One of the discoveries made during the early studies of brain scans conducted by Professors Antoine Lutz and Richard Davidson (in which I participated) was that meditation on nonreferential compassion—a meditation practice based on the union of emptiness and compassion—produced a profound increase in what are often referred to as *gamma waves,* fluctuations in the electrical activity of the brain measured by EEG scans, that reflect an integration of information among a wide variety of brain regions.[6] A gamma wave is a very high frequency brain wave, often associated with attention, perception, consciousness, and the kind of neuronal synchrony discussed in Part One. Many neuroscientists understand gamma waves as representing activity that occurs when various neurons communicate in a spontaneously synchronous manner across large areas of the brain.

Preliminary research indicates that long-term meditation practitioners spontaneously exhibit high levels of gamma wave activity, suggesting that the brain achieves a more stable and integrated state during meditation. Because neuroscience and the technology available for study are still relatively new, however, we can't definitively say that meditation practice increases communication across wider areas of the brain. Nevertheless, the study of London cabdrivers mentioned earlier seems to suggest that repeated experience does change the structure of the brain—which implies that focusing on the transparency of thoughts, emotions, and sensory experiences may very well transform related areas of the brain.

THE FRUIT OF COMPASSION

> *Even small merit done brings great happiness.*
> —The Collection of Meaningful Expressions,
> translated by Elizabeth M. Callahan

As mentioned before, calm-abiding meditation is like charging your mental and emotional batteries. Compassion is the mental and emotional "technology" that uses the recharged batteries in a proper way. What I mean here by a "proper way" is that there's always the possibility that you might misuse the abilities you've developed through shinay meditation just to enhance your own mental and emotional stability to gain power over, or even harm, others. After you've gained some experience, though, compassion and shinay meditation are normally practiced together.

When you join compassion meditation with shinay practice, you benefit not only yourself but others as well. Real progress on the path includes an awareness of benefiting yourself and others simultaneously.

Compassion is reciprocal. As you develop your own mental and emotional stability and extend that stability through a compassionate understanding of others and dealing with them in a kind, empathetic way, your own intentions or aspirations will be fulfilled more quickly and easily. Why? Because if you treat others compassionately—with the understanding that they have the same desire for happiness and the same desire to avoid unhappiness that you do—then the people around you feel a sense of attraction, a sense of wanting to help you as much as you help them. They listen more closely to you, and develop a sense of trust and respect. People who might once have been adversaries begin to treat you with more respect and consideration, facilitating your own progress in completing difficult tasks. Conflicts resolve themselves more easily, and you'll find yourself advancing more quickly in your career, beginning new relationships without the usual heartaches, and even starting a family or improving your existing family relationships more easily—all because you've charged your batteries through shinay meditation and extended that charge through developing a kinder, more understanding, and more empathetic relationship with others. In a sense, compassion practice demonstrates the truth of interdependence in action. The more openhearted you become toward others, the more openhearted they become toward you.

When compassion begins to awaken in your own heart, you're able to be more honest with yourself. If you make a mistake, you can acknowledge it and take steps to correct it. At the same time, you're less likely to look for flaws in other people. If people do something offensive, if they start screaming at you or treating you badly, you'll notice (probably with some surprise) that you don't react in the same way you once might have.

A woman I met a couple of years ago while I was teaching in Europe approached me to describe a problem she was having with her neighbor. Their cottages were quite close together, separated only by their own very narrow gardens. It seemed the neighbor was always trying to annoy her in small ways, by tossing things in her yard, damaging her plants, and so on. When she asked him why he was doing these things, he replied, "I love annoying people."

Of course, as these petty attacks continued, the woman became very angry and found herself unable to resist retaliating in the same petty ways. Gradually the "garden wars" grew fiercer, and the animosity between the two neighbors increased.

Clearly frustrated, the woman asked me what she should do to solve the problem so that she could go about her life peacefully. I advised her to meditate on compassion for her neighbor.

"I tried that already," she replied. "It didn't work."

After talking with her a bit about how she had practiced, I explained that meditating on compassion involves more than trying to invoke a sense of warmth or kindness for someone we find irritating or frustrating. It actually requires a bit of analytical investigation into the other person's motivations, as well as an attempt to develop some sense of understanding of the other person's feelings—an understanding that, just like ourselves, everyone shares the same basic desire to be happy and to avoid unhappiness.

When I returned to Europe the following year, she approached me again, this time smiling very happily. She told me that everything had changed. When I asked her what had happened, she explained, "I practiced the way we talked a year ago, thinking about what my neighbor felt and what his motivation might be—how he just wanted to be happy and avoid unhappiness just as I did. And after a while I suddenly realized I wasn't afraid of him anymore. I realized that nothing he did could hurt me. Of course he kept on trying, but nothing he did really bothered me anymore. It was as if, by meditating on compassion for him, I developed confidence in myself. I didn't have to retaliate or get angry, because whatever he was doing seemed pretty harmless and small.

"After a while," she continued, "he started to become embarrassed. Once he realized that nothing he did was going to get me to respond, not only did he stop trying to annoy me, he actually became quite shy every time he saw me—and eventually he went from being very shy to very polite. One day he came to me and apologized for doing all those annoying things. In a way, I think, it seemed that by meditating on compassion for him, as I became confident in myself, he gradually developed confidence in himself as well. He didn't have to do anything to prove how powerful or damaging he could be."

Most of us don't live in isolation. We live in an interdependent world. If you want to improve the condition of your own life, then you need to depend on others to help you along the way. Without this kind of interdependent relationship you would not have food, a roof over your head, or a job to go to—you wouldn't even be able to buy coffee from Starbucks! So, if you deal with others in a compassionate, empathetic way, you can only improve the conditions of your own life.

When you look at your relationship to the world and your own life in this way, you see that loving-kindness and compassion are very, very powerful.

The other great benefit of developing compassion is that through understanding the needs, fears, and desires of others, you develop a deeper capacity to understand your own self—what you hope for, what you hope to avoid, and the truth about your own nature. And this, in turn, serves to dissolve whatever sense of loneliness or low self-esteem you may be feeling. As you begin to recognize that everyone craves happiness and is terrified of unhappiness, you start to realize that you're not alone in your fears, needs, or desires. And in realizing this, you lose your fear of others—everyone is a potential friend, a potential brother or sister—because you share the same fears, the same longings, and the same goals. And with this understanding, it becomes so much easier to really communicate with others on a heart-to-heart level.

One of the best examples of this kind of openhearted communication was related to me by a Tibetan friend who is a taxi driver in New York City. One day he made a wrong turn—crossing the wrong way down a one-way street in the middle of a rush of traffic. A police officer stopped him and gave him a ticket and a summons to appear in court. When he appeared in court, one of the people in line in front of him was very angry, shouting at the judge, the officer who'd issued his ticket, and the lawyers around him. His outrageous behavior didn't gain him much sympathy from the court; he lost his case and ended up having to pay a large fine.

When my friend's turn came to appear before the judge, he relaxed and smiled and said a kind good morning to the police officer who had issued the ticket, politely asking how he was doing that day. At first the officer was a little taken aback. But then he replied, "Hi. I'm fine. How are you?" My friend greeted the judge in the same polite manner. As the court proceedings began, the judge asked my friend, "So why did you make that wrong turn?" My friend explained—again, very politely—that the traffic was so bad that day, he didn't have any other choice. The judge turned to the police officer and asked him if the account was true, and the officer admitted that the traffic had been very bad that day, and that the mistake my friend had committed was understandable under the circumstances. So the judge dismissed the charge and let my friend go. Afterward, in the lobby, the officer came up to my friend and said, "You did very well."

For my friend—and for me as well—that court experience served as a nice example of the benefits of practicing simple kindness and compassion, of treating people as you yourself would want to be treated, and not as adversaries. No matter what your position in life—whether you're a taxi driver, a powerful politician, or a high-level corporate executive—your chances of happiness are greatly increased by treating whomever you're dealing with as a friend, someone who has the same hopes and fears as you. The effect of this approach is exponential. If you can only affect the attitude or outlook of one person, that one person will be able to transmit the effects of that change to another. If you can shift the attitude of three people, and each of those people can shift the attitude of three more people, you've changed the lives of twelve people. And the chain reaction just grows and grows.

17

THE BIOLOGY OF HAPPINESS

Arouse confidence in the principle of cause and effect from the depths of your heart.

—Patrul Rinpoche, *The Words of My Perfect Teacher,*
translated by the Padmakara Translation Group

A REALLY GOOD scientific experiment produces as many questions as it does answers. And one of the big questions generated by the study of trained meditators has been whether their ability to direct their minds results from factors like similar genetic makeup, shared cultural and environmental backgrounds, or similarities in the way they were trained. In other words, can ordinary people, who weren't taught from childhood in the specialized environment of a Tibetan Buddhist monastery, benefit from practicing any of the techniques of Buddhist meditation?

Because the clinical research involving Buddhist meditation masters is still in its infancy, it may be a long time before we can answer such questions with real assurance. It can be said, however, that the Buddha taught hundreds, probably thousands, of ordinary people—farmers, shepherds, kings, businessmen, soldiers, beggars, and even common criminals—how to direct their minds in ways that would create the kinds of subtle changes in their physiology that would allow them to override their biological and environmental conditioning and achieve a lasting state of happiness. If what he'd taught hadn't been effective, no one would know his name, there would be no tradition known as Buddhism, and you wouldn't be holding this book in your hands.

ACCEPTING YOUR POTENTIAL

Whatever is the cause that binds is the path that liberates.
—The Ninth Gyalwang Karmapa,
Mahāmudrā: The Ocean of Definitive Meaning,
translated by Elizabeth M. Callahan

You don't need to have been a particularly nice person to be able to start the "inside job" of being happy. One of the greatest Tibetan Buddhist masters of all time was a murderer. Now he's considered a saint, and paintings of him always show him with a hand cupped to his ear, listening to the prayers of ordinary people.

His name was Milarepa. The only child of a wealthy couple, he was born sometime in the tenth century C.E. When his father died unexpectedly, his uncle took control of the family's wealth and forced Milarepa and his mother to live in poverty, a change in circumstance that wasn't accepted very enthusiastically by either of them. None of their other relatives spoke up for them; it was simply the fate of widows and children of that time to accept the decisions made by the men of the family.

As the story goes, when Milarepa came of age, his mother sent him to study with a sorcerer, so that he could learn some dark spell to take revenge on his relatives. Fueled both by his anger and by a desire to please his mother, Milarepa mastered the art of dark magic, and on the day of his cousin's wedding he cast a spell that caused his uncle's house to collapse, killing thirty-five of his family members in one blow.

Whether Milarepa actually used magic or some other means to kill his family can be debated. The fact remains that he killed his relatives and was afterward filled with a terrible feeling of guilt and remorse. If telling a single lie to one person can keep you awake at night, imagine how murdering thirty-five members of your own family would make you feel.

To atone for his crime, Milarepa left his home to devote his life to the welfare of others. He traveled to southern Tibet to study under a man named Marpa, who'd made three separate trips to India to collect the essence of the Buddha's teachings in order to bring it back to

Tibet. In most respects, Marpa was an ordinary person—a "house-holder" in Buddhist terms, which meant that he had a wife and children, owned a farm, and was occupied by the daily concerns of running his business and coping with his family. But he was also devoted to the Dharma, and his devotion gave him great courage. Walking across the Himalayas from Tibet to India isn't an easy task, and most people who try it die in the attempt. His timing was extraordinary, though, because not long after his final journey, India was conquered by invaders and all the Buddhist libraries and monasteries were destroyed, while most of the monks and teachers who'd perpetuated the Buddha's training were killed.

Marpa had passed all the knowledge he'd brought back from India to his eldest son, Dharma Dode. But Dharma Dode was killed in a riding accident, and even as he was recovering from his loss, Marpa sought an heir to the teachings he'd received in India. He took one look at Milarepa and saw in him a man who had what it took not only to master the details of the teachings, but also to grasp the very essence of them and pass it on to the next generation. Why? Because Milarepa's heart was completely broken over what he'd done, and the depth of his remorse was so great that he was willing to go to any lengths to make amends.

Through experience alone Milarepa had come to recognize one of the most basic of the Buddha's teachings: Everything you think, everything you say, and everything you do is reflected back to you as your own experience. If you cause someone pain, you experience pain ten times worse. If you promote others' happiness and well-being, you experience the same happiness ten times over. If your own mind is calm, then the people around you will experience a similar degree of calmness.

This understanding has been around for a long time, and has been expressed in different ways by different cultures. Even Heisenberg's famous uncertainty principle acknowledges an intimate connection between inner experience and physical manifestation. The really exciting development for our time is that modern technology has begun to enable researchers to demonstrate the principle in action. Today's researchers are starting to provide objective evidence that learning to

calm the mind and developing a more compassionate attitude produces higher levels of personal pleasure, and can actually change the
structure and function of the brain in ways that ensure that happiness
remains constant over time.

In order to test the effects of Buddhist meditation practice on
ordinary individuals, Richard Davidson and his colleagues designed a
study involving employees at a Midwestern corporation.[1] His goal was
to determine whether the techniques could help offset the psychological and physical effects of workplace stress. He invited employees
at the corporation to sign up for a course in meditation, and after
performing some initial blood work and EEG tests, randomly divided
the participants into two groups: one that would immediately be
trained and a control group that would receive the training after the effects on the first group had been thoroughly studied. The training in
meditation was given over a period of ten weeks by Dr. Jon Kabat-
Zinn, professor of medicine at the University of Massachusetts and
founder of the Stress Reduction Clinic at the University of Massachusetts Memorial Medical Center.

Continuing to evaluate the subjects of the study for several months
after they'd completed their meditation training, Davidson and his
team found that within three or four months after the training ended,
EEG tests began to show a gradual and significant increase in electrical activity in the left prefrontal lobe area, the region of the brain associated with positive emotions. During the same three- or four-month
period, the subjects of the study themselves began reporting experiences of reduced stress, greater calmness, and a more general sense of
well-being.

But an even more interesting result was about to be discovered.

HAPPY MIND, HEALTHY BODY

A human being's exceptional physical, verbal, and mental en-
dowment provides the unique ability to pursue a constructive
course of action.

—JAMGÖN KONGTRUL, *The Torch of Certainty,*
translated by Judith Hanson

There's been very little disagreement between Buddhists and modern scientists that a person's state of mind has some effect on the body. To use an everyday example, if you've had a fight with someone during the day or received a notice in the mail that your electricity is about to be shut off because you haven't paid your bill, chances are you won't be able to sleep well when you go to bed. Or if you're about to make a business presentation or talk to your boss about a problem you're having, your muscles might tense up, you might feel sick to your stomach, or you might suddenly develop a pounding headache.

Until recently, there wasn't a lot of scientific evidence to support the connection between a person's state of mind and his or her physical experience. Richard Davidson's study of corporate employees had been carefully designed so that the end of the meditation training would coincide with the annual flu shots provided by their company. After resampling the blood work of the subjects involved in the study, he found that the people who'd received meditation training showed a significantly higher level of influenza antibodies than those who hadn't been trained. In other words, people who'd demonstrated a measurable shift in left prefrontal lobe activity also showed an enhancement in their immune systems.

Results of this kind represent a huge advance in modern science. Many of the scientists I've talked to have long suspected that there is a connection between the mind and the body. But prior to this study, evidence of the connection had not yet been so clearly indicated.[2]

During its long and remarkable history, science has focused almost exclusively on looking at what goes *wrong* with the mind and body rather than at what goes *right*. But there's been a slight shift in the wind recently, and now it appears that many people in the modern scientific community are being offered the chance to look more closely at the anatomy and physiology of happy, healthy human beings.

Within the past several years, a number of projects have demonstrated very strong links between positive mental states and a reduction in the risk or intensity of various physical illnesses. For example, Dr. Laura D. Kubzansky, Assistant Professor, Department of Society, Human Development, and Health at the Harvard School of Public Health, initiated a study that followed the medical histories of about 1,300 men over a period of ten years.[3] The subjects of the study were primarily military veterans, who had access to a level of medical care that many people don't enjoy, so their medical histories were fairly complete and easy to track over such a long period. Because "happiness" and "unhappiness" are somewhat broad terms, for the purposes of the study Dr. Kubzansky focused on specific manifestations of these emotions: namely, optimism and pessimism. These characteristics are defined by a standardized personality test[4] that equates optimism with the belief that your future will be satisfying because you can exercise some control over the outcome of important events, and pessimism with the belief that whatever problems you're experiencing are unavoidable because you have no control over your destiny.

At the end of the study, Dr. Kubzansky found that after statistically adjusting for factors including age, gender, socioeconomic status, exercise, alcohol intake, and smoking, the incidence of some forms of heart disease among subjects identified as optimists was nearly 50 percent less than that of subjects identified as pessimists. "I'm an optimist," Dr. Kubzansky said in a recent interview, "but I didn't expect results like this."[5]

Another research study, led by Dr. Laura Smart Richman, Assistant Research Professor of Psychology, Duke University, looked at the physical effects of two other positive emotions associated with happiness: hopefulness and curiosity.[6] Nearly 1,050 patients of a multipractice clinic agreed to participate by responding to a questionnaire about their emotional states, physical behaviors, and other information such as income and educational level.

Dr. Richman and her team tracked the medical records of these patients over the course of two years. Again, after statistically adjusting for the contributing factors mentioned above, Dr. Richman found that higher levels of hope and curiosity were associated with a lower likelihood of either having or developing diabetes, high blood pressure, and respiratory tract infections. In the typically careful scientific language aimed at downplaying sensational claims, Dr. Richman's study concluded that the results suggested that "positive emotion may play a protective role in the development of disease."[7]

THE BIOLOGY OF BLISS

The support is the supreme, precious human body.

—GAMPOPA, *The Jewel Ornament of Liberation,*
translated by Khenpo Konchog Gyaltsen Rinpoche

The funny thing about the mind is that if you ask a question and then listen quietly, the answer usually appears. So I don't doubt that the development of the technology capable of examining the mind's effect on the body has something to do with modern scientists' growing interest in studying the mind-body relationship. So far, the questions asked by scientists have been quite reasonably cautious, and the answers they've received have been provocative but not overwhelmingly conclusive. Because the scientific study of happiness and its attributes is still relatively young, we have to allow for some uncertainty. We have to give it the time to go through its growing pains.

Meanwhile, scientists have begun making connections that may be able to help provide objective explanations for the effectiveness of Buddhist training. For example, the blood samples Richard Davidson took from the subjects of his study showed that people who demonstrated the type of prefrontal lobe activity associated with positive emotion also evidenced lower levels of cortisol, a hormone naturally produced by the adrenal glands in response to stress.[8] Because cortisol tends to suppress the function of the immune system, some correlation can be made between feeling more or less confident, happy, and able to exert some control over one's life, and having a stronger, healthier immune system. By contrast, a general sense of being unhappy, out of control, or dependent on external circumstances tends to produce higher levels of cortisol, which in turn can weaken the immune system and make us more vulnerable to all sorts of physical diseases.

THE BENEFITS OF RECOGNIZING EMPTINESS

> *You yourself become a living teaching; you yourself become living dharma.*
>
> —CHÖGYAM TRUNGPA, *Illusion's Game*

Any of the meditation practices described in Part Two can help to alleviate the sense of being "out of control" through the patient observation of the thoughts, emotions, and sensations we experience in any given moment, from which comes a gradual recognition that they are not inherently real things. If every thought or feeling you experienced was an inherently real thing, your brain would probably be crushed by the sheer weight of their accumulation!

"Through practice," a student of mine once said, "I've learned that feelings are not facts. They come and go depending on my own state of restlessness or calm at any given time. If they were facts, they wouldn't change, regardless of my own situation."

The same can be said of thoughts, perceptions, and physical sensations, all of which are, according to Buddhist teachings, momentary expressions of the infinite possibility of emptiness. They're like people moving through an airport on their way to a different city. If you asked them their intentions, they'd tell you that they were "just passing through."

So how can recognizing emptiness reduce the levels of stress that contribute to physical disease? Earlier we looked at the ways in which emptiness might be compared to our experiences in dreams, using the specific example of a car. The car we experience in a dream isn't "real" in the conventional sense of being made up of various material parts assembled in a factory; nevertheless, as long as our dream lasts, our dream experience of driving around in the car seems very real. We enjoy "real" pleasure in driving the car and showing it off to our friends and neighbors, and experience "real" unhappiness if we get involved in an accident. But the car in the dream doesn't truly exist, does it? It's only because we're caught up in the deep ignorance of dreaming that whatever we experience while driving the car appears real.

Yet, even in dreams, certain conventions reinforce our acceptance of the reality of dream experiences. For example, when we dream about a waterfall, in general the water falls downward. If we dream about fire, the flames reach upward. When our dreams turn into nightmares—for example, if we get involved in a car accident, find ourselves having to jump from a tall building and crash to the ground, or are compelled to walk through fire—the suffering we experience in the dream seems very real.

So let me ask a question that may be a little bit harder to answer than some of the others I've asked along the way: What method could you use to free yourself from that kind of suffering in the dream state without waking up?

I've asked this question many times in public teachings and received a number of different answers. Some of the responses are very funny, like the suggestion from one person who proposed hiring a clairvoyant housekeeper who would instinctively recognize your pain and step into the dream and guide you through the difficulties. I'm not sure how many clairvoyant housekeepers are available for hire, or whether their chances of being hired would improve by listing clairvoyance as a special skill on their résumés.

Other people have suggested that spending time meditating in the waking state will automatically improve one's chances of having more pleasant dreams. Unfortunately, I can't say that I've ever found this to be the case among the people I've met and spoken with around the world. Still others have suggested that if you dream about jumping off a building, you might suddenly be able to discover that you can fly. I don't know how or why this could happen, but it seems a rather risky proposition.

Very rarely, someone will suggest that the best possible solution is to recognize in the dream that you're merely dreaming. As far as I've learned, that is the best answer. If you can recognize while caught up in a dream that you're only dreaming, you can do anything you like within the dream. You can jump off a tall building without being hurt. You can jump into a fire without being burned. You can walk on water without drowning. And if you're driving your dream car and have an accident, you can escape unharmed.

The point, though, is that by training in recognizing the emptiness of all phenomena, you can accomplish amazing things in waking life. Most people go through waking life caught up in the same delusions of limitation and entrapment they experience in their dreams. But if you spend even a few minutes out of every day examining your thoughts and perceptions, you'll gradually gain the confidence and awareness of recognizing that your everyday experience isn't as solid or unalterable as you once thought it was. The neuronal gossip you once accepted as truth will gradually begin to shift, and the communication between your brain cells and the cells associated with your senses will change accordingly. Bear in mind that the change will almost always occur very slowly. You have to give yourself a chance to let the transformation take place in its own time, according to your own nature. If you try to rush the process, at best you'll be disappointed; at worst, you could hurt yourself (for example, I wouldn't advise trying to walk through fire after only a couple of days meditating on emptiness).

I can't think of a better example of the patience and diligence required to really recognize your full potential, your Buddha nature, than the first in the series of *Matrix* movies, which many of you probably saw years before I did. The movie impressed me not only because the conventional reality experienced by people caught up in the Matrix was eventually revealed as an illusion, but also because even with the benefit of all the equipment and training available to him, it still took the main character, Neo, a while to recognize that the personal limitations he'd accepted as real for most of his life were in fact only projections of his own mind. When he first had to confront these limitations, he was scared, and I could easily identify with his fear. Even though he had Morpheus as his guide and teacher, he still found it hard to believe in what he was truly capable of—just as I found it hard to believe in the truth of my own nature when it was first revealed to me by masters who'd actually demonstrated the full potential of their true nature. Only at the end of the movie, when Neo had to experience on his own that the lessons he'd been taught were true, was he able to stop bullets in midair, fly through space, and see things before they actually occurred.

Still, he had to learn these things in a gradual way. So don't expect that after two or three days of meditation you'll be able to walk on water or fly off buildings. More than likely, the first change you'll notice is a greater degree of openness, confidence, and self-honesty, and an ability to recognize the thoughts and motivations of other people around you more quickly than you might previously have been able to do. That's no small accomplishment; it's the beginning of wisdom.

If you keep practicing, all the wonderful qualities of your true nature will gradually reveal themselves. You'll recognize that your essential nature cannot be harmed or destroyed. You'll learn to "read" the thoughts and motivations of others even before they understand them themselves. You'll be able to look more clearly into the future and see the consequences of your own actions and the actions of people around you. And, perhaps most important of all, you'll realize that in spite of your own fears, no matter what happens to your physical body, your true nature is essentially indestructible.

18

MOVING ON

Consider the advantages of this rare human existence.
—JAMGÖN KONGTRUL, *The Torch of Certainty,*
translated by Judith Hanson

AMONG ALL LIVING creatures studied thus far by modern scientists, only human beings can be said with absolute certainty to have been endowed with the ability to make deliberate choices about the direction of their lives, and to discern whether those choices will lead them through the valley of transitory happiness or into a realm of a lasting peace and well-being. Though we may be genetically wired for temporary happiness, we've also been gifted with the ability to recognize within ourselves a more profound and lasting sense of confidence, peace, and well-being. Among sentient beings, human beings appear to stand alone in their ability to recognize the necessity to forge a bond between reason, emotion, and the instinct to survive, and in so doing create a universe— not only for themselves and the human generations that follow, but also for all creatures who feel pain, fear, and suffering—in which we all are able to coexist contentedly and peaceably.

This universe already exists, even if we don't realize it at present. The aim of Buddhist teachings is to develop the capacity to recognize that this universe—which is really nothing more or less than the infinite possibility inherent within our own being—exists in the here and now. In order to recognize it, however, it is necessary to learn how to rest the mind. Only through resting the mind in its natural awareness can we begin to recognize that we are not our thoughts, not our feelings, and not our perceptions. Thoughts, feelings, and perceptions are functions of the body. And everything I've learned as a Buddhist and everything I've learned about modern science tells me that human beings are more than just their bodies.

The exercises I've presented in this book represent only the first stage of the path toward realization of your full potential, your Buddha nature. On their own, these exercises about learning to calm your mind, becoming familiar with it, and developing a sense of loving-kindness and compassion can effect undreamed-of changes in your life. Who wouldn't want to feel confident and calm in the face of difficulties, reduce or eliminate their sense of isolation, or contribute, however indirectly, to the happiness and well-being of others, providing thereby an environment in which we ourselves, those we love and care for, and generations as yet unborn can flourish? All it takes to accomplish these marvels is a little patience, a little diligence, a little willingness to let go of conditioned ideas about yourself and the world around you. All it takes is a bit of practice in waking up in the middle of the dreamscape of your life and recognizing that there is no difference between the experience of the dream and the mind of the dreamer.

Just as the landscape of a dream is infinite in scope, so is your Buddha nature. The stories surrounding Buddhist masters of the past are full of wonderful tales of men and women who walked on water, passed through fire unharmed, and communicated telepathically with their followers across great distances. My own father was able to undergo the experience of a surgeon slicing through the sensitive layers of skin and muscle around his eye without feeling pain.

I can also share with you a few interesting stories about a man who lived in the twentieth century who achieved his full potential as a sentient being. That man was the Sixteenth Karmapa, the previous head of the Kagyu lineage of Tibetan Buddhism. In the wake of the difficulties that shook Tibet in the late 1950s, he and a large group of followers resettled in Sikkim, in northern India, where he founded a large monastery, several schools, and a variety of institutions to support a thriving community for exiled Tibetans. Once the community in Sikkim was securely established, the Karmapa began traveling the world, teaching the growing number of people who at that time were just beginning to become aware of the special nature of Tibetan Buddhism. In the course of his travels through Europe and North America, he performed what might be described as miracles, such as leaving his footprints in solid rock, and bringing rain to drought-stricken areas of the American Southwest—on one occasion causing a spring to appear spontaneously in a desert region occupied by Hopi Indians.

But it was the manner of the Sixteenth Karmapa's death that offered those who witnessed it the most vivid demonstration of the qualities of natural mind. In 1981 he was treated for cancer at a hospital outside of Chicago. The course of his illness bewildered his medical team, as his symptoms seemed to come and go for no apparent reason, disappearing altogether at times, only to reappear later in some previously unaffected area of his body—as though, according to one description, "his body were joking with the machines."[1] Throughout the ordeal, the Karmapa never complained of pain. He was much more interested in the well-being of the hospital staff, many of whom stopped by regularly simply to experience the enormous sense of tranquillity and compassion that radiated from him despite the ravages of disease.

When he died, the lamas and other Tibetans who'd stayed with him throughout his treatment asked that his body remain undisturbed for three days, as is the Tibetan custom after the passage of a great master. Because the Karmapa had made such a profound impression on the hospital staff, the administration granted their request, and, rather than immediately removing his remains to the hospital morgue, they allowed his body to remain in his room, seated in the meditation posture in which he'd died.

As documented by the doctors who examined him over the course of those three days, the Karmapa's body never underwent rigor mortis, and the area around his heart remained nearly as warm as that of a living person. More than twenty years later, the condition of his body after death defies medical explanation, and still leaves a profound impact on those who witnessed it.

I suspect that his decision to be treated and to leave his body in a Western hospital was the Sixteenth Karmapa's last, and perhaps greatest, gift to humanity: a demonstration to the Western scientific community that we do indeed possess capacities that cannot be explained in ordinary terms.

FINDING A TEACHER

You must be guided by an authentic spiritual mentor.
—The Ninth Gyalwang Karmapa,
Mahāmudrā: The Ocean of Definitive Meaning,
translated by Elizabeth M. Callahan

The interesting aspect regarding the masters of the past and present is that they shared a similar process of training. They began by practicing many of the exercises in calming the mind and developing compassion presented in this book, and then reached their full potential by following the lead of a teacher wiser and more experienced than themselves. If you want to go further, if you want to explore and experience your full potential, you need a guide. You need a teacher.

What are the qualities of a good teacher? First of all, the teacher must have been trained according to a lineage—otherwise, he or she may just be making up the rules or guidelines of practice out of his or her own pride, or perpetuating a misunderstanding of what he or she has read in books. There is also a great but subtle power in receiving guidance from a teacher trained in an established lineage tradition: the power of interdependence discussed in Part One. When you work with a teacher trained in a lineage, you become part of the "family" of that lineage. Just as you learned unspoken yet invaluable lessons from your birth family or the family in which you were raised, you will gain priceless lessons just through observing and interacting with a true lineage teacher.

In addition to having been trained in the disciplines of a particular lineage, a qualified teacher must also demonstrate compassion and, through his or her actions, subtly make clear his or her own realization without ever mentioning it. Avoid teachers who talk about their own accomplishments—because that kind of talk or boasting is a sure sign that they have not achieved realization at all. Teachers who have had some experience never speak about their own accomplishments, but tend, instead, to speak about the qualities of their own teachers. And yet you can sense their own qualities through the aura of authority that envelops them, like the light reflecting from a nugget of gold. You don't see the gold itself, but only the brilliance of golden light.

CHOOSING HAPPINESS

> *Intention is the karma of the mind.*
>
> —GUNAPRABHA, *The Treasury of Abhidharma,*
> translated by Elizabeth M. Callahan

Just watch a child playing a video game, obsessed with pushing buttons to kill enemies and win points, and you'll see how addictive such games can be. Then take a step back and see how the financial, romantic, or other "games" you've been playing as an adult are just as addictive. The main difference between an adult and a child is that an adult has the experience and understanding to step away from the game. An adult can choose to look more objectively at his or her mind, and in doing so develop a sense of compassion for others who haven't been able to make that choice.

As described throughout the preceding pages, once you commit yourself to developing an awareness of your Buddha nature, you'll inevitably start to see changes in your day-to-day experience. Things that used to trouble you gradually lose their power to upset you. You'll become intuitively wiser, more relaxed, and more openhearted. You'll begin to recognize obstacles as opportunities for further growth. And as your illusory sense of limitation and vulnerability gradually fades away, you'll discover deep within yourself the true grandeur of who and what you are.

Best of all, as you start to see your own potential, you'll also begin to recognize it in everyone around you. Buddha nature is not a special quality available to a privileged few. The true mark of recognizing your Buddha nature is to realize how ordinary it really is—the ability to see that every living creature shares it, though not everyone recognizes it in themselves. So instead of closing your heart to people who yell at you or act in some other harmful way, you find yourself becoming more open. You recognize that they aren't just jerks, but are people who, like you, want to be happy and peaceful; they're only acting like jerks because they haven't recognized their true nature and are overwhelmed by sensations of vulnerability and fear.

Your practice can begin with the simple aspiration to do better, to approach all of your activities with a greater sense of mindfulness, and to open your heart more deeply toward others. Motivation is the single most important factor in determining whether your experience is conditioned by suffering or by peace. Mindfulness and compassion actually develop at the same pace. The more mindful you become, the easier you'll find it to be compassionate. And the more you open your heart to others, the more mindful you become in all your activities.

At any given moment, you can choose to follow the chain of thoughts, emotions, and sensations that reinforce a perception of yourself as vulnerable and limited, or to remember that your true nature is pure, unconditioned, and incapable of being harmed. You can remain in the sleep of ignorance, or remember that you are and always have been awake. Either way, you're still expressing the unlimited nature of your true being. Ignorance, vulnerability, fear, anger, and desire are expressions of the infinite potential of your Buddha nature. There's nothing inherently wrong or right with making such choices. The fruit of Buddhist practice is simply the recognition that these and other mental afflictions are nothing more or less than choices available to us because our real nature is infinite in scope.

We choose ignorance because we *can*. We choose awareness because we *can*. Samsara and nirvana are simply different points of view based on the choices we make in how to examine and understand our experience. There's nothing magical about nirvana and nothing bad or wrong about samsara. If you're determined to think of yourself as limited, fearful, vulnerable, or scarred by past experience, know only that you have *chosen* to do so, and that the opportunity to experience yourself differently is always available.

In essence, the Buddhist path offers a choice between familiarity and practicality. There is, without question, a certain comfort and stability in maintaining familiar patterns of thought and behavior.

Stepping outside that zone of comfort and familiarity necessarily involves moving into a realm of unfamiliar experience that may seem really scary—an uncomfortable in-between realm like the one I experienced in retreat. You don't know whether to go back to what was familiar but frightening or to forge ahead toward what may be frightening simply because it's unfamiliar.

In a sense, the uncertainty surrounding the choice to recognize your full potential is similar to what several of my students have told me about ending an abusive relationship: There's a certain reluctance or sense of failure associated with letting go of the relationship. The primary difference between severing an abusive relationship and entering the path of Buddhist practice is that when you enter the path of Buddhist practice you're ending an abusive relationship with yourself. When you choose to recognize your true potential, you gradually begin to find yourself belittling yourself less frequently, your opinion of yourself becomes more positive and wholesome, and your sense of confidence and the sheer joy of being alive increases. At the same time, you begin to recognize that everyone around you has the same potential, whether they know it or not. Instead of dealing with them as threats or adversaries, you'll find yourself able to recognize and empathize with their fear and unhappiness and spontaneously respond to them in ways that emphasize solutions rather than problems.

Ultimately, happiness comes down to choosing between the discomfort of becoming aware of your mental afflictions and the discomfort of being ruled by them. I can't promise you that it will always be pleasant to simply rest in the awareness of your thoughts, feelings, and perceptions, and to recognize them as interactive creations between your own mind and body. In fact, I can pretty much guarantee that looking at yourself this way will be, at times, extremely unpleasant. But the same can be said about beginning anything new, whether it's going to the gym, starting a job, or beginning a diet.

The first few months are always difficult. It's hard to learn all the skills you need to master a job; it's hard to motivate yourself to exercise; it's hard to eat healthfully every day. But after a while the difficulties subside, you start to feel a sense of pleasure or accomplishment, and your entire sense of self begins to change.

Meditation works the same way. For the first few days you might feel very good, but after a week or so, practice becomes a trial. You can't find the time, sitting is uncomfortable, you can't focus, or you just get tired. You hit a wall, as runners do when they try to add an extra half mile to their exercise. The body says, "I can't," while the mind says, "I should." Neither voice is particularly pleasant; in fact, they're both a bit demanding.

Buddhism is often referred to as the "middle way" because it offers a third option. If you just can't focus on a sound or a candle flame for one second longer, then by all means stop. Otherwise, meditation becomes a chore. You'll end up thinking, *Oh no, it's 7:15. I have to sit down and cultivate happiness.* No one ever progresses that way. On the other hand, if you think you could go on for another minute or two, by all means do so. You may be surprised by what you learn. You might discover a particular thought or feeling behind your resistance that you didn't want to acknowledge. Or you may simply find that you can actually rest your mind longer than you thought you could—and that discovery alone can give you greater confidence in yourself, while at the same time reducing your level of cortisol, increasing your level of dopamine, and generating more activity in the left prefrontal lobe of your brain. And these biological changes can make a huge difference in your day, providing a physical reference point for calmness, steadiness, and confidence.

But the best part of all is that no matter how long you meditate, or what technique you use, every technique of Buddhist meditation ultimately generates compassion, whether we're aware of it or not. Whenever you look at your mind, you can't help but recognize your similarity to those around you. When you see your own desire to be happy, you can't avoid seeing the same desire in others, and when you look clearly at your own fear, anger, or aversion, you can't help but see that everyone around you feels the same fear, anger, and aversion. When you look at your own mind, all the imaginary differences between yourself and others automatically dissolve and the ancient prayer of the Four

Immeasurables becomes as natural and persistent as your own heart-beat:

> *May all sentient beings have happiness and the causes of*
> *happiness.*
> *May all sentient beings be free from suffering and the causes*
> *of suffering.*
> *May all sentient beings have joy and the causes of joy.*
> *May all sentient beings remain in great equanimity, free from*
> *attachment and aversion.*

NOTES

Introduction

1. An annual public dialogue between Buddhists and modern scientists initiated in 1987 by the Dalai Lama and Francesco Varela, one of the foremost neuroscientists of the twentieth century.

2. Michael D. Lemonick, "The Biology of Joy," *Time,* January 17, 2005.

Chapter 2. The Inner Symphony

1. Jeremy W. Hayward and Francisco J. Varela, *Gentle Bridges: Conversations with the Dalai Lama on the Science of Mind* (Boston: Shambhala, 1992), 188.

2. Daniel Goleman, *Emotional Intelligence* (New York: Bantam Books, 1995), 15.

Chapter 5. The Relativity of Perception

1. Jeremy W. Hayward and Francisco J. Varela, *Gentle Bridges: Conversations with the Dalai Lama on the Science of Mind* (Boston: Shambhala, 1992), 183–84.

2. Ibid., 199.

Chapter 7. Compassion: Survival of the Kindest

1. *Mind & Life XI,* DVD-ROM 4 (Boulder: Mind and Life Institute, 2003).

2. Ibid.

Chapter 8. Why Are We Unhappy?

1. See, for example, Thomas Lewis, M.D., Fari Amini, M.D., and Richard Lannon, M.D., *A General Theory of Love* (New York: Random House, 2000), 68–99.

2. Philip Brinkman, "Lottery Winners and Accident Victims: Is Happiness Relative?" *Journal of Personality and Social Psychology* 36 (1978): 917.

Chapter 13. Compassion: Opening the Heart of the Mind

1. Jamgön Kongtrul, *The Torch of Certainty*, translated by Judith Hanson (Boston: Shambhala, 1977), 60–61.

Chapter 16. An Inside Job

1. Daniel Goleman, *Destructive Emotions: How Can We Overcome Them?* (New York: Bantam Dell, 2003), 194–95.

2. Ibid., 4–27.

3. For more information, see the essay "Homo homini lupus? Morality, the Social Instincts, and Our Fellow Primates," in J.-P. Changeux, A. R. Damasio, W. Singer, and Y. Christen, eds., *Neurobiology of Human Values* (Heidelberg: Springer-Verlag, 2005).

4. Thomas Lewis, M.D., Fari Amini, M.D., and Richard Lannon, M.D., *A General Theory of Love* (New York: Random House, 2000), 62ff.

5. Ibid., 176ff.

6. See A. Lutz et al, "Long-Term Meditators Self-Induce High-Amplitude Gamma Synchrony During Mental Practice," *Proceedings of the National Academy of Science* 101 (2004): 16369–73.

Chapter 17. The Biology of Happiness

1. For complete details, see R. Davidson et al, "Alterations in Brain and Immune Function Produced by Mindfulness Meditation," *Psychosomatic Medicine* 65 (2004): 564–70.

2. Daniel Goleman, *Destructive Emotions: How Can We Overcome Them?* (New York: Bantam Dell, 2003), 334.

3. For complete details, see L. Kubzansky et al, "Is the Glass Half Empty or Half Full? A Prospective Study of Optimism and Coronary Heart Disease in the Normative Aging Study," *Psychosomatic Medicine* 63 (2001): 910–16.

4. The Minnesota Multiphasic Personality Inventory.

5. Michael D. Lemonick, "The Biology of Joy," *Time,* January 17, 2005.

6. L. Richman et al, "Positive Emotion and Health: Going Beyond the Negative," *Health Psychology* 24, no. 4 (2005): 422–29.

7. Ibid.

8. Lemonick, "Biology of Joy."

Chapter 18. Moving On

1. Ken Holmes, *Karmapa* (Forres, Scotland: Altea Publishing, 1995), 32.

GLOSSARY

Absolute bodhicitta Direct insight into the nature of mind. *See also* Application bodhicitta, Aspiration bodhicitta, Bodhicitta, Relative bodhicitta.

Absolute reality The infinite potential for anything to occur. *See also* Emptiness, Tongpa-nyi.

Action potential The actual transmission of a signal between one neuron and another.

Amygdala A neuronal structure in the brain involved in forming the emotional aspects of memory, particularly fear and pleasure.

Application bodhicitta Taking steps to cultivate the liberation of all sentient beings from all forms and causes of suffering through recognition of their Buddha nature. *See also* Absolute bodhicitta, Aspiration bodhicitta, Bodhicitta, Relative bodhicitta.

Aspiration bodhicitta Cultivation of the heartfelt desire to raise all sentient beings to the level at which they recognize their Buddha nature. *See also* Absolute bodhicitta, Application bodhicitta, Bodhicitta, Relative bodhicitta.

Autonomic nervous system The area of the brain stem that automatically regulates muscle, cardiac, and glandular responses.

Axon The trunk of a nerve cell.

Bodhicitta Sanskrit: The "mind" or "heart" of awakening. *See also* Absolute bodhicitta, Application bodhicitta, Aspiration bodhicitta, Relative bodhicitta.

Body The physical aspect of existence. *See also* Mind, Speech.

Brain stem The lowest and oldest layer of the human brain, responsible for regulating involuntary functions such as metabolism, heart rate, and the fight-or-flight response. *See also* Reptilian brain.

Buddha nature The natural state of all sentient beings, which is infinitely aware, infinitely compassionate, and infinitely able to manifest itself. *See also* Enlightenment, Natural mind.

Clarity Spontaneous awareness; the unlimited cognizant aspect of the mind. Also known as the clear light of mind.

Dharma Sanskrit: The truth, or the way things are.

Dul-tren Tibetan: Smallest particle.

Dul-tren-cha-may Tibetan: Indivisible particle.

Dzinpa Tibetan: Grasping or fixation.

Dzogchen Tibetan: The Great Perfection.

Electron An electronically charged subatomic particle.

Emptiness The inherently indescribable basis of all phenomena from which anything and everything arises. *See also* Absolute reality, Tongpa-nyi.

Enlightenment In Buddhist terms, the firm and unshakable recognition of one's Buddha nature. *See also* Buddha nature, Natural mind.

Four Noble Truths, The The name applied to the first teachings given by the Buddha in Varanasi after he attained enlightenment; also known as the first of the Three Turnings of the Wheel of Dharma.

Gewa Tibetan: An adjective used to describe something that empowers or strengthens; often translated as "virtuous."

Gom Tibetan: Literally, "to become familiar with"; the common term for meditation.

Heart Sons The main students of a major teacher.

Hippocampus A neuronal structure in the brain involved in forming the verbal and spatial aspects of memory.

Hypothalamus A neuronal structure at the base of the limbic region involved in the process of releasing hormones into the bloodstream.

Interdependence The coming together of different causes and conditions to create a specific experience.

Kagyu A Tibetan Buddhist lineage based on the oral transmission of teachings from master to student; from the Tibetan words *ka,* meaning "speech," and *gyu,* meaning "lineage."

Karma Sanskrit: Action or activity.

Karmapa The head of the Karma Kagyu lineage of Tibetan Buddhism.

Le-su-rung-wa Tibetan: Pliability.

Limbic region The middle layer of the brain, which includes neuronal connections that provide the capacity to experience emotions and the impulse to nurture.

Limbic resonance A kind of brain-to-brain capacity to recognize the emotional states of others through facial expression, pheromones, and body or muscular position.

Limbic revision The capacity to change or revise the neuronal circuitry of the limbic region through direct experience with another person.

Loving-kindness In Buddhist philosophical terms, the aspiration that all other sentient beings—even those we dislike— experience the same sense of joy and freedom that we ourselves aspire to feel.

Magakpa Tibetan: Unimpededness; often translated as "ability" or "power." The aspect of Buddha nature that transcends habitual ideas of personal limitation.

Mahamudra Sanskrit: Great Seal or Great Gesture.

Mahasiddha Sanskrit: A person who has passed through extraordinary trials to achieve profound understanding.

Mala Sanskrit: A string of prayer beads, usually used to count repetitions of a mantra.

Mantra Sanskrit: The repetition of special combinations of ancient syllables.

Mass The measure of the amount of matter in an object.

Mi-gewa Tibetan: An adjective used to describe something that weakens; often translated as "nonvirtuous."

Mind The aspect of existence that involves consciousness. *See also* Body, Speech.

Mindfulness Resting the mind in bare awareness of thoughts, feelings, and sensory experiences.

Natural mind The mind in its natural state, free from conceptual limitations. *See also* Buddha Nature, Enlightenment.

Neocortex The uppermost layer of the brain, specific to mammals, which provides capacities for reasoning, forming concepts, planning, and fine-tuning emotional responses.

Neuron A nerve cell.

Neuronal plasticity The capacity to replace old neuronal connections with new ones.

Neuronal synchrony A process in which neurons move across widely separated areas of the brain spontaneously and instantaneously communicate with one another.

Neurotransmitter A substance that passes electrochemical signals among neurons.

Nirvana Sanskrit: Extinguishing or blowing out (as in the blowing out of the flame of a candle); often interpreted as the state of total bliss or happiness arising from the extinguishing or "blowing out" of the ego or the idea of "self."

Nying-jay Tibetan: Compassion; an utterly direct expansion of the heart.

Nyingma A Tibetan term roughly translated as "old ones"; refers specifically to the oldest lineage of Tibetan Buddhism established in Tibet during the seventh century C.E.

Photon A particle of light.

Purusha Sanskrit: Literally, "something that possesses power"; usually used to refer to a human being.

Relative bodhicitta The intention, within the relativistic framework of self and other, to raise all sentient beings to the level at which they recognize their Buddha nature. *See also* Absolute bodhicitta, Application bodhicitta, Aspiration bodhicitta, Bodhicitta.

Relative reality The moment-by-moment experience of endless changes and shifts of thoughts, emotions, and sensory perceptions.

Reptilian brain The lowest and oldest layer of the human brain, responsible for regulating many involuntary functions, such as metabolism, heart rate, and the fight-or-flight response. *See also* Brain stem.

Samaya Sanskrit: A vow or commitment.

Samsara Sanskrit: Wheel; in Buddhist terms, the wheel of suffering.

Sautantrika Sanskrit: An early school of Buddhist philosophy.

Sem Tibetan: That which knows.

Sentient being Any creature endowed with the capacity to think or feel.

Shamata Sanskrit: Calm abiding practice; simply allowing the mind to rest calmly as it is. *See also* Shinay.

Shastra Sanskrit: An explanation of or commentary on an idea or term expressed by the Buddha during his lifetime.

Shedra Tibetan: A monastic college.

Shinay Tibetan: Calm abiding; simply allowing the mind to rest calmly as it is. *See also* Shamata.

Spectrum The set of energy levels, which is different for each type of atom.

Speech The aspect of existence that involves verbal and nonverbal communication. *See also* Mind, Body.

Sutra Sanskrit: Literally, "thread." In Buddhist terminology, a specific reference to the actual words of the Buddha "threaded" throughout the years of his teachings.

Synapse The gap across which neurons communicate.

Tathagatagarbha Sanskrit: "The nature of one who has gone that way," a way of describing someone who has attained complete enlightenment; also translated as "Buddha nature," "enlightened essence," "ordinary nature," and "natural mind."

Thalamus A neuronal structure located at the very center of the brain, through which sensory messages are sorted before being passed to other areas of the brain.

Three Turnings of the Wheel of Dharma The three sets of teachings on the nature of experience given by the Buddha at different times and places.

Tonglen Tibetan: "Sending and taking." The practice of sending all one's happiness to other sentient beings and taking in their suffering.

Tongpa Tibetan: Indescribable, inconceivable, unnamable, empty of meaning in ordinary terms.

Tongpa-nyi Tibetan: *See* Emptiness.

Tulku Tibetan: An enlightened master who has chosen to reincarnate in human form.

Vaibhasika Sanskrit: An early school of Buddhist philosophy.

Velocity The speed and direction of the movement of subatomic particles.

SELECT BIBLIOGRAPHY

Dhammapada, The. Translated by Eknath Easwaran. Tomales, Calif.: Nilgiri Press, 1985.

Gampopa. *The Jewel Ornament of Liberation.* Translated by Khenpo Konchog Gyaltsen Rinpoche. Edited by Ani K. Trinlay Chödron. Ithaca, N.Y.: Snow Lion Publications, 1998.

Goleman, Daniel. *Emotional Intelligence.* New York: Bantam Books, 1995.

———. *Destructive Emotions: How Can We Overcome Them?* New York: Bantam Dell, 2003.

Hayward, Jeremy W., and Francisco J. Varela. *Gentle Bridges: Conversations with the Dalai Lama on the Science of Mind.* Boston: Shambhala, 1992.

Kongtrul, Jamgön. *The Torch of Certainty.* Translated by Judith Hanson. Boston: Shambhala, 1977.

Lewis, Thomas, M.D., Fari Amini, M.D., and Richard Lannon, M.D. *A General Theory of Love.* New York: Random House, 2000.

Patrul Rinpoche. *The Words of My Perfect Teacher,* rev. ed. Translated by the Padmakara Translation Group. Boston: Shambhala, 1998.

Śāntideva. *The Bodhicaryāvatāra.* Translated by Kate Crosby and Andrew Skilton. New York: Oxford University Press, 1995.

Tsoknyi Rinpoche. *Carefree Dignity.* Compiled and translated by Erick Pema Kunsang and Marcia Binder Schmidt. Edited by Kerry Morgan. Kathmandu: Rangjung Yeshe Publications, 1998.

Twelfth Khenting Tai Situpa, The. *Ground, Path, and Fruition.* Auckland: Zhyisil Chokyi Ghatsal Charitable Trust Publications, 2005.

Tulku Urgyen Rinpoche. *As It Is.* Vol. 1. Translated by Erik Pema Kunsang. Compiled by Marcia Binder Schmidt. Edited by Kerry Morgan. Hong Kong: Ranjung Yeshe Publications, 1999.

ACKNOWLEDGMENTS

FOR THEIR INSPIRATION and instruction, I would like to thank every one of my teachers, including H. E. Tai Situ Rinpoche; H. H. Dilgo Khyentse Rinpoche; Saljay Rinpoche; Nyoshul Khen Rinpoche; my father, Tulku Urgyen Rinpoche; Khenchen Kunga Wangchuk Rinpoche; Khenpo Losang Tenzin; Khenpo Tsultrim Namdak; Khenpo Tashi Gyaltsen; Drupon Lama Tsultrim; my grandfather Tashi Dorje; and Dr. Francisco J. Varela.

For their unfailing help in providing scientific information and clarification, I would also like to thank Dr. Richard Davidson, Dr. Antoine Lutz, Dr. Alfred Shapere, Dr. Fred Cooper, Dr. Laura D. Kubzansky, Dr. Laura Smart Richman, and C. P. Antonia Sumbundu. I would also like to thank Anne Benson, Ani Jamdron, Dr. Alex Campbell, Christian Bruyat, Lama Chodrak, Edwin and Myoshin Kelley, Dr. E. E. Ho, Dr. Felix Moos, Helen Tvorkov, Jacqui Horne, Jane Austin Harris, Jill Satterfield, M. L. Harrison Mackie, Veronique Tomaszewski-Ramses, and Dr. William Rathje for their contributions in proofreading and editing the manuscript.

This manuscript would not have come to light without the help of my agent, Emma Sweeney; my publisher, Shaye Areheart; my editor, John Glusman; and the support of Tim and Glenna Olmsted, Mei Yen and Dwayne Ladle, Robert Miller, Christine Mignanelli, Lama Karma Chotso, Dr. Elaine Puleo, Gary Swanson, and Nancy Swanson.

I would like to offer very special thanks to my brother, Tsoknyi Rinpoche; Josh Baran; Daniel Goleman; Tara Bennett-Goleman; and Lama Tashi, for their inspiration, kindness, and generosity in organizing all the many aspects of this book.

Finally, I would like to thank Eric Swanson, who worked with great patience in spite of my constant changes to the manuscript. He is very knowledgeable and has an open mind and a constant smile. Without his tremendous effort, this book would not have been completed.

INDEX

ABOUT THE AUTHORS

YONGEY MINGYUR RINPOCHE

Born in 1975 in Nubri, Nepal, Yongey Mingyur Rinpoche is a rising star among the new generation of Tibetan Buddhist masters trained outside of Tibet. Deeply versed in the practical and philosophical disciplines of the ancient tradition of Tibetan Buddhism, he is also conversant in the practical and theoretical issues and details of modern culture, having taught for nearly a decade around the world, meeting and speaking with a diverse array of renowned scientists and ordinary people yearning to rise above the suffering inherent in the human condition and achieve a state of lasting happiness. His honest, often humorous accounts of his own personal difficulties have endeared him to thousands of Buddhist and non-Buddhist students across the world. For more information about Mingyur Rinpoche, his teachings, and his activities, please visit the Yongey Foundation website at www.mingyur.org.

ERIC SWANSON

A graduate of Yale University and the Julliard School, Eric Swanson formally adopted Buddhism in 1995. He is the author of *What the Lotus Said,* a graphic description of his journey to Tibet as a member of a team of volunteers developing schools and medical clinics in rural areas occupied predominantly by nomadic populations, and coauthor of *Karmapa: The Sacred Prophecy,* a history of the Karma Kagyu lineage of Tibetan Buddhism.